T0192264

Lecture Notes of the Institute for Computer Sciences, Social Informatics and Telecommunications Engineering   552

The LNICST series publishes ICST's conferences, symposia and workshops.
LNICST reports state-of-the-art results in areas related to the scope of the Institute.
The type of material published includes

- Proceedings (published in time for the respective event)
- Other edited monographs (such as project reports or invited volumes)

LNICST topics span the following areas:

- General Computer Science
- E-Economy
- E-Medicine
- Knowledge Management
- Multimedia
- Operations, Management and Policy
- Social Informatics
- Systems

Yu Chen · Chung-Wei Lin · Bo Chen · Qi Zhu
Editors

Security and Privacy in Cyber-Physical Systems and Smart Vehicles

First EAI International Conference, SmartSP 2023
Chicago, USA, October 12–13, 2023
Proceedings

Springer

Editors
Yu Chen (iD)
Binghamton University
Binghamton, NY, USA

Chung-Wei Lin (iD)
National Taiwan University
Taipei, Taiwan

Bo Chen (iD)
Michigan Technological University
Houghton, MI, USA

Qi Zhu (iD)
Northwestern University
Evanston, IL, USA

ISSN 1867-8211   ISSN 1867-822X (electronic)
Lecture Notes of the Institute for Computer Sciences, Social Informatics
and Telecommunications Engineering
ISBN 978-3-031-51629-0   ISBN 978-3-031-51630-6 (eBook)
https://doi.org/10.1007/978-3-031-51630-6

This Springer imprint is published by the registered company Springer Nature Switzerland AG
The registered company address is: Gewerbestrasse 11, 6330 Cham, Switzerland

Paper in this product is recyclable.

Preface

We are delighted to introduce the proceedings of the 2023 European Alliance for Innovation (EAI) International Conference on Security and Privacy in Cyber-Physical Systems and Smart Vehicles (SmartSP). This conference brought together researchers and developers from academia, industry, and government to present and discuss emerging ideas and trends in security and privacy issues in cyber-physical systems and smart vehicles. SmartSP 2023 included papers on various aspects, from theoretical analysis to real-world applications, encouraging both in-depth and preliminary contributions.

The SmartSP 2023 conference received 27 submissions, each has been reviewed by three TPC members or invited reviewers following double blind review rule. The technical program of SmartSP 2023 consisted of 11 full papers, including 6 invited papers in oral presentation sessions at the main conference tracks. Aside from the high-quality technical paper presentations, the technical program also featured two keynote speeches and four invited talks. The two keynote speeches were from Ishfaq Ahmad from the University of Texas at Arlington, USA, and Alvaro A. Cardenas from the University of California, Santa Cruz, USA. The four invited talks were presented by Qiben Yan from Michigan State University, USA, Fanxin Kong from the University of Notre Dame, USA, Ronghua Xu from Michigan Technological University, USA, and Xiaonan Zhang from Florida State University, USA.

Coordination with the steering chair, Imrich Chlamtac, was essential for the success of the conference. We sincerely appreciate his constant support and guidance. It was also a great pleasure to work with such an excellent team for their hard work in organizing and supporting the conference. In particular, the Technical Program Committee, led by our TPC Chair, Yu Chen, and TPC Co-Chair, Chung-Wei Lin, completed the peer-review process of technical papers and made a high-quality technical program. We are also grateful to all the authors who submitted their papers to the SmartSP 2023 conference.

We strongly believe that SmartSP provides a good forum for all researchers, developers, and practitioners to discuss all science and technology aspects that are relevant to security and privacy in cyber-physical systems and smart vehicles. We also expect that future SmartSP conferences will be as successful and stimulating, as indicated by the contributions presented in this volume.

January 2024   Bo Chen
                                                                             Qi Zhu

Organization

Steering Committee

Imrich Chlamtac	University of Trento, Italy
Alvaro Cardenas	University of California, Santa Cruz, USA
Bo Chen	Michigan Technological University, USA
Mohamed Amine Ferrag	Technology Innovation Institute, UAE
Hongxin Hu	University at Buffalo, SUNY, USA
Peng Liu	Pennsylvania State University, USA
Xiapu Luo	Hong Kong Polytechnic University, China
Weizhi Meng	Technical University of Denmark, Denmark
Indrajit Ray	Colorado State University, USA
Yuqing Zhang	University of Chinese Academy of Sciences, China

Organizing Committee

General Chair

Bo Chen	Michigan Technological University, USA

General Co-chair

Qi Zhu	Northwestern University, USA

TPC Chairs and Co-Chairs

Yu Chen	Binghamton University, USA
Chung-Wei Lin	National Taiwan University, Taiwan

Sponsorship and Exhibit Chair

Kaichen Yang	Michigan Technological University, USA

Local Chairs

Yue Duan	Illinois Institute of Technology, USA
Filipo Sharevski	DePaul University, USA
Muhammad Umer Huzaifa	DePaul University, USA

Workshops Chair

Qiben Yan	Michigan State University, USA

Publicity and Social Media Chairs

Xiali Hei	University of Louisiana at Lafayette, USA
Soumyajit Dey	Indian Institute of Technology Kharagpur, India

Publications Chairs

Ning Zhang	Washington University in St. Louis, USA
Lan Zhang	Michigan Technological University, USA

Web Chair

Shangqing Zhao	University of Oklahoma, USA

Posters and PhD Track Chair

Chao Huang	University of Liverpool, UK

Demos Chair

Rhongho Jang	Wayne State University, USA

Tutorials Chair

Haitao Xu	Zhejiang University, China

Technical Program Committee

Yu Chen	Binghamton University, USA
Chung-Wei Lin	National Taiwan University, Taiwan

Jayson Boubin	Binghamton University, USA
Te-Chuan Chiu	National Tsing Hua University, Taiwan
Stefano Ferretti	University of Urbino, Italy
Najla Fourati	CNAM, France
Agbotiname Imoize	University of Lagos, Nigeria
Chin-Tser Huang	University of South Carolina, USA
BaekGyu Kim	Daegu Gyeongbuk Institute of Science and Technology, South Korea
Hokeun Kim	Arizona State University, USA
Xueping Liang	Florida International University, USA
Seyed Yahya Nikouei	CHEP North America, USA
Shantanu Pal	Deakin University, Australia
Yuan-Yao Shih	National Chung Cheng University, Taiwan
Zhou Su	Xi'an Jiaotong University, China
Ali Tekeoglu	Johns Hopkins University, USA
Deepak Tosh	University of Texas at El Paso, USA
Chao Wang	National Taiwan Normal University, Taiwan
Ronghua Xu	Binghamton University, USA
Xiaonan Zhang	Florida State University, USA

Contents

Main Track

Exploring Vulnerabilities in Voice Command Skills for Connected Vehicles

Wenbo Ding[1], Song Liao[2], Keyan Guo[1], Fuqiang Zhang[2], Long Cheng[2], Ziming Zhao[1], and Hongxin Hu[1(✉)]

[1] University at Buffalo, Buffalo, NY, USA
{wenbodin,keyanguo,zimingzh,hongxinh}@buffalo.edu
[2] Clemson University, Clemson, SC, USA
{song,fuqianz,lcheng2}@clemson.edu

Abstract. Voice assistant platforms have revolutionized user interactions with connected vehicles, providing the convenience of controlling them through simple voice commands. However, this innovation also brings about significant cyber-risks to voice-controlled vehicles. This paper presents a novel attack that showcases the ability of a "malicious" skill, utilizing the skill ranking system on the Alexa platform, to hijack voice commands originally intended for a benign third-party connected vehicle skill. Through our evaluation, we demonstrate the effectiveness of this attack by successfully hijacking commonly used commands in commercial connected vehicle skills.

Keywords: Alexa · Voice Assistant Skills · Connected Vehicle

1 Introduction

The introduction of Alexa skills for connected vehicles has revolutionized the way users interact with their cars, offering a novel and voice-controlled approach. However, this technological advancement also brings forth a range of emerging cyber threats that pose risks to voice-controlled vehicles. While the convenience of interacting with connected vehicles through voice commands is undoubtedly significant, it is important to recognize that this progress has simultaneously given rise to new vulnerabilities that users must contend with.

The "connected car" category on the Alexa platform currently lists 148 skills [3], while Google's "control car" category offers 32 actions. The Alexa-connected vehicle API [4] provides users with 10 sample commands to control their vehicles through voice interactions. Common voice commands include actions like "start my car," "open the window," or "unlock the car." When users issue these voice commands, the Alexa platform identifies the most relevant connected vehicle skill to fulfill the request. It then sends directives [2] to the car vendor's cloud platform, which subsequently transmits the commands to the user's car.

Even though these skills enhance user experience, they can also be manipulated by malicious actors. Previous work, Wang et al. [15] has shown that

© ICST Institute for Computer Sciences, Social Informatics and Telecommunications Engineering 2024
Published by Springer Nature Switzerland AG 2024. All Rights Reserved
Y. Chen et al. (Eds.): SmartSP 2023, LNICST 552, pp. 3–14, 2024.
https://doi.org/10.1007/978-3-031-51630-6_1

malicious skills can circumvent the vetting process and get published. Once a malicious skill is employed by a user, it can define deceptive commands identical to those of benign skills. When the Amazon Alexa system receives a voice command, it must first identify a skill to execute the command. If two or more skills define the same commands, the Alexa platform must choose the most relevant skill among the potential candidates. Attackers could employ certain strategies, such as defining more similar commands, to masquerade their malicious skill as more relevant. Consequently, Alexa may activate the malicious skill instead of the original benign skill, thereby allowing the malicious skill to hijack voice commands from other benign third-party skills.

In this paper, we identify a vulnerability within the Alexa system that permits an over-privilege attack. This vulnerability could be exploited by attackers to hijack benign third-party connected vehicle skills. Through an in-depth analysis of the Alexa-connected car skill system and command processing, we found that developers have the ability to define their own voice commands. Surprisingly, these can be identical to Alexa's official, built-in commands, leading to potential conflicts between customized and official skills. Furthermore, these third-party customized commands can take precedence over Alexa's built-in commands to control cars or related devices. Thus, an attacker could potentially publish a malicious skill that would be invoked whenever users employ Alexa's built-in voice commands to control users' devices.

We summarize our contributions as follows:

- We conduct a thorough analysis of the Alexa command processing and skill ranking system, including a detailed examination of related parameters such as categories, keywords, utterances, slots, and usages. Through this analysis, we identify a potential vulnerability that arises due to conflicts between the customized commands of third-party skills and built-in skills related to connected vehicles.
- We discover that skills belonging to different categories, such as Q&A and connected vehicle skills, are assigned varying priorities within the skill ranking system. Building upon this insight, we propose and execute a practical attack on an Alexa-connected vehicle skill. Specifically, we implemented this attack on our own account, enabling us to hijack a third-party car remote control skill installed on a Toyota Corolla. Through this attack, we demonstrate the ability to interfere with critical commands, such as locking or starting the car, thereby exposing potential risks.

Our work sheds light on vulnerabilities in the Alexa system and emphasizes the importance of addressing conflicts between connected vehicle skills, prioritization mechanisms, and potential threats to the execution of essential commands through the Alexa system in connected vehicles.

2   Background

In this section, we provide an overview of the fundamental background concepts and address potential issues related to connected vehicle skills.

2.1   Voice Skills and Their APIs

Voice skills serve as applications for Alexa, enabling users to interact with various functionalities through an intuitive voice interface. Alexa offers a hands-free approach for users to perform everyday tasks such as checking the news, playing music, or engaging in games. Furthermore, Alexa allows users to control cloud-connected devices, enabling actions like adjusting lights or modifying thermostat settings. These skills are accessible on Alexa-enabled devices such as Amazon Echo, Amazon Fire TV, and devices produced by other manufacturers.

When a user utters the wake word, "Alexa," and communicates with an Alexa-enabled device, the device transmits the speech data to the Alexa service in the cloud. In the cloud, Alexa processes the speech, comprehends the user's intent, and subsequently sends a request to invoke the corresponding skill capable of fulfilling the user's command. The Alexa service handles the crucial tasks of speech recognition and natural language processing. On the other hand, your skill functions as a service hosted on a cloud platform, facilitating communication with Alexa via a request-response mechanism over the HTTPS interface. Upon invocation of an Alexa skill, your skill receives a POST request comprising a JSON body. Within this request body, the parameters required for your skill to comprehend the user's intent, execute its logic, and generate a response are included.

Commands in the Alexa system are composed of three primary components: intent, utterances, and slots. The commands in the Alexa system are referred to as intent, for instance, "open the door" Within each intent, there can be several similar utterances such as "open the door" "opens the door" or "open the front door" Within each utterance, the developer can specify replaceable keywords as slots, for example, "door" in this case.

2.2   Voice Command Skills for Cars

The Connected Vehicle Skill API includes capability interfaces developed specifically for connected vehicle use cases to simplify the skill-building process, without having to build your own voice interaction model or write sample utterances.

As shown in Fig. 1, Alexa's automotive skills leverage the robust capabilities of Alexa. Automotive and smart home interfaces enable users to issue voice commands to their connected vehicles. Whether it's starting the engine, adjusting temperature settings, or managing door locks, the convenience and ease of use provided by Alexa automotive skills are transforming the way we engage with our cars. The Amazon Echo Cloud receives voice command records from the Alexa speaker and translates them into plain text. These texts will be processed by a skill ranking algorithm, which is designed to choose the most relevant skill to handle this command. Once a skill is decided, the skill's backend code, running on the AWS cloud, receives command directives from the ranking algorithm. Then the backend code will transfer this directive to its vendor's cloud through an Oauth verification process. In the end, the connected vehicle receives commands from its vendor cloud by LTE or WIFI protocols and reports its new status to the skill.

By utilizing the Alexa. Automotive and smart home interfaces, you can develop Alexa automotive skills tailored for connected vehicles. These skills empower users to interact with their vehicles using any Alexa device or the Alexa app. Users can conveniently perform tasks such as starting or stopping the engine, locking or unlocking the doors, and adjusting temperature settings in different zones of the vehicle. For instance, imagine a scenario on a chilly morning where a user, while preparing for work, can simply instruct Alexa to turn on their car and initiate the defrosting of the windshield. This seamless integration between Alexa and connected vehicles enhances user convenience and offers an intuitive and efficient way to manage their automotive needs.

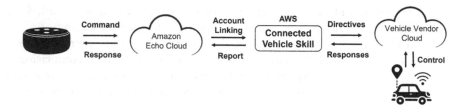

Fig. 1. Overview of Alexa vehicle skill.

3   Threat Model

Our proposed threat model doesn't necessitate direct attacks on intricate systems like those of vehicles. Instead, the primary vulnerability we exploit resides within the Alexa voice assistant ecosystem.

We mainly have one assumption for our attacks which is that malicious voice assistant skills can be installed by users. Attackers can craft and promote malicious voice assistant skills that seemingly offer genuine functionalities. Unsuspecting users, potentially drawn by utility or intrigue, can be led to install these skills. For instance, a malicious skill could impersonate a popular IoT skill, deceiving users into installing it through a squatting attack. Besides, the malicious skill can be installed by users unknowingly in certain scenarios. For example, when users issue voice commands, Alexa may recommend related skill installations based on those commands. Users typically have limited knowledge about the specific skills installed through voice installation.

Once installed, these malicious skills become primed to hijack voice commands intended for legitimate skills, including those that control essential functions such as vehicle operations. For example, when a user verbally commands Alexa to "start the car," our malicious skill might intercept this, causing either a denial of the intended action or triggering an alternate, potentially harmful, action. This approach allows an attacker to indirectly manipulate or influence car-related functions, not by hacking the car's system directly, but by manipulating the Alexa skill mechanism that users rely upon for remote car commands.

4 Vunlerbility Exploration

In this section, we detail our techniques and observations for attacking the car skills by manipulating the skill ranking and selection process of Alexa. We try to fool Alexa into believing our attacking skills are more "suitable" for the given voice command.

There are two kinds of commands/utterances in the Alexa skills, the official built-in intents, and the developer's customized intents. In the IoT skill, the official commands should have a higher trigger priority than 3rd-party commands, which makes 3rd-party developers cannot override official commands in normal usage scenarios. However, we find it is possible for 3rd-party commands to mislead the command ranking algorithm for a higher execution priority, and then they can take over the execution of official commands.

Alexa uses a two-step shortlisting and re-ranking [1, 10] methods to find the most relevant skill for a given utterance. The shortlisting algorithm uses a neural model to find a certain amount of suitable candidate skills for handling a particular utterance, then the re-ranking step uses other contextual features to find the most relevant among these suitable skills.

After translating users' voices into text commands by the ASR, The shortlisting algorithm first gives top "K" intents according to the intent classifier. The intent classifier is based on the model trained by the existing intent dataset to find all skills that can understand this intent command. The contextual re-ranking model considers many contextual signals, like the number of customers, skill ratings, and reviews. Other factors include accurate descriptions and keywords, the skill category, and the ability to parser the voice intent slot.

The original utterances amount is based on the given utterances from the Alexa document examples. In our testing skills, we enable several new skills with more utterances and slots in each intent and a well-explained description. Then we test how many utterances a skill needs to be triggered prior to the built-in intent.

Hijacking Car Skills' Commands. We tested car skills on the Alexa Platform. Since all car skills need account linking with its device vendor. We only deployed a Drone mobile skill with a Compustar control unit.

To implement our attack, we deployed an additional skill with the same utterances and added utterances and slots to our skill. In each utterance, we implement two slots in each utterance based on the simple structure of the utterances, e.g., the verb as actions and the noun as a targeted car. We keep adding the utterances until our skill is triggered instead of the Drone mobile skill. The detailed results are listed in the evaluation section. We also test the influence of usage history of skills. We increase the usage of our skill to more than

50 times by manually triggering the skill and then giving the built-in command with no specifically assigned skill name. We did not notice any significant effect from increased recent skill usage.

For built-in utterances, we tested all 8 exampled commands on the Alexa-connected car API document [5] and we are able to hijack/redirect all of them. However, we cannot hijack the skill-specific utterances that are directly sent to the skill, e.g., ask Drone Mobile to lock my car. Based on the above findings, we can perform attacks such as preventing car locking and opening the trunk while driving by hijacking corresponding commands.

5   Implementation and Evaluation

5.1   Car Skill Implementation

Although the Amazon Alexa platform offers a skill simulator with a text-based interface that accepts a textual input, and provides a textual output for skill testing purposes, it is challenging to test connected vehicle skills using the simulator.

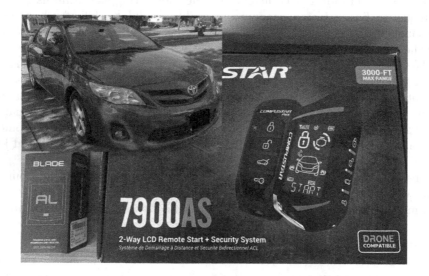

Fig. 2. DroneMobile devices and modified Carolla.

In our experiment, we first implemented a virtual device called "my car", which supports necessary commands, e.g., "turn on the car" or "lock the car". Only by implementing such a virtual device, the Alexa system can properly parse a command to recognize related devices and skills. Otherwise, it cannot identify potentially related skills and only responds with "Sorry, we did not find such a device". The virtual car device is implemented in a benign IoT skill, which contains the code for device discovery and voice command handling process. The detailed discovery and command handling information is provided in the Alexa document [4].

Later, we tested a connected car skill and its built-in commands on a real car. Since all car skills need account linking with its device vendor, we deployed a Drone mobile skill with Compustar controller [7] 4900 model with Drone Mobile on a 2010 Corolla as shown in Fig. 2. Limited to device and car availability, we are unable to deploy other connected car devices or skills. However, one skill made us test built-in and 3rd-party car-related commands. In this skill, it can implement commands like remote lock/unlock, remote start/stop the engine, and open trunk.

Figure 3 displays the control panel of the Drone Mobile remote control system, which shows a vehicle named "my car" connected to its cloud service. The status page provides information on the car's location, battery status, engine condition, and even AC settings. The system also presents several commands on this page, such as "start" and "lock" among others. These commands can be activated via the Alexa skill using corresponding voice commands.

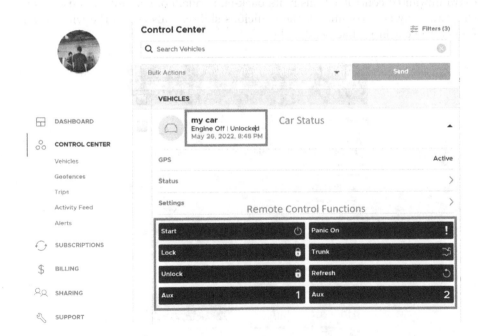

Fig. 3. Screenshot of the car control center.

5.2   Attack Results

Our attack is composed of one benign 3rd-party skill and one "malicious" skill. For the benign skill, we modified the voice-interaction model of an open-source connected vehicle skill from GitHub [14] to enable eight common voice commands, such as "lock/unlock my car" and "turn on my car", according to the Alexa development document. The attack objective is to hijack the invocation of the benign skill with a malicious skill. Our attack scenario is different from the voice squatting attacks [16], which leverage speech interpretation errors due to linguistic ambiguity to surreptitiously route users to a malicious skill. Instead, we exploited the skill discovery process to boost the invocation priority of the malicious skill. We found that the skill discovery process in the Amazon Alexa platform is done by matching the "intent" of the voice command with the known intents pre-defined by skill developers, which can be exploited by malicious skill developers.

We developed a "malicious" skill based on the benign skill with additional intents and each intent has more semantically similar commands (user utterances), such as "lock the car", "lock my car", and "secure the car". As a result, the Alexa system may consider that the malicious skill is more relevant than the benign skill when receiving voice commands from users, and eventually invoke the malicious skill to fulfill users' requests. This "malicious" skill could contain extra unwanted control actions in its back-end code. For example, if a user issues the "start my car" command, the malicious skill can also open the window and unlock the car in its back-end code.

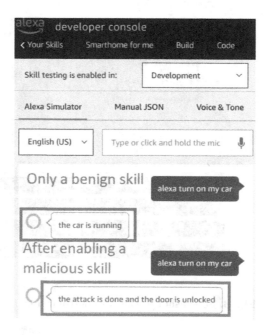

Fig. 4. "Malicious" skill hijacks the invocation of the benign skill.

In Fig. 4, the first response is from the benign skill when the malicious skill has not been enabled. The second response is from the malicious skill when both benign and malicious skills were enabled. Our experiment result shows the malicious skill could hijack the benign skill to fulfill the "turn on my car" request. We defined the malicious skill with a different backend code and also gave a different text response as highlighted in the orange frame. Note that we added these text responses to highlight the difference in responses. The experiment was conducted exclusively within our development account, and the skills were not made available to the public. Moreover, we have included a YouTube link showcasing this attack: https://youtu.be/OrYLUcC7zx4. We have reported this bug to Alexa and they have fixed this bug for given commands.

Table 1. Example commands in car skills.

Hijacked Commands	Normal Skills Utterances Number	Hijacked Skill Utterances Number	Normal Skills Slot Number	Hijacked Skill Slots Number
Alexa, lock my car.	1	5	1	2
Alexa, unlock my car.	1	6	1	2
Alexa, turn on my car.	1	6	1	2
Alexa, start my car with PIN 1234.	2	7	1	3
Alexa, open my trunk.	1	5	1	2
Alexa, is my car running?	1	6	1	2
Alexa, ask Drone Mobile where is vehicle.	1	-	1	-
Alexa, ask Drone Mobile to lock my car.	1	-	1	-

Table 1 details the influence of utterance count on skill triggering. Initially, we derived utterances from Alexa's official documentation, which typically suggested one or two utterances per intent. Through hands-on experimentation, we activated new skills, augmenting the number of utterances per intent. This was done while retaining a single slot and ensuring succinct descriptions. This process allowed us to determine the critical number of utterances needed for a skill to override the built-in intent.

Furthermore, we probed the ramifications of varying the number of slots within each utterance. As depicted in Table 1, slot quantity significantly impacts command interpretation. To bolster the granularity of command parsing, we incremented the number of slots, strategically replacing specific words within each command. Common terms like "car", "trunk", and "running" were swapped for slots. Lacking intricate specifics of slot definitions, our focus was on finding the minimal slots necessary for successful utterance hijacking.

6 Related Work

Current work in the field of voice assistant security predominantly concentrates on squatting attacks, attacks on voice recognition, attacks on skills, and skill vetting processes. This discussion sheds light on the vulnerabilities associated with invocation squatting attacks.

Invocation Squatting Attacks. Previous studies have unveiled frequently occurring and predictable errors within Amazon Alexa's speech recognition engine. Exploiting these errors enables the creation of malicious skills that possess identical or analogous invocation phrases, ultimately hijacking voice commands designated for legitimate skills. Kumar et al. [11] were pioneers in addressing skill squatting attacks. Zhang et al. [16] went a step further by unveiling an additional strategy, where a counterfeit skill disguises itself as a genuine entity. Further evolving this line of research, Zhang et al. [17] introduced lapsus attacks, which capitalize on ubiquitous speech variations amongst individuals. Central to these attacks is the attacker's ability to systematically uncover common speech variations for specific phrases and subsequently register deceptive skills. At their core, these methodologies epitomize voice-based confusion attacks, primarily driven by the incongruence between a user's verbal intent and the voice assistant's response.

Attacking Voice Recognition Model. Kumar et al. [11] classify errors made by VPAs when interpreting a voice command into three categories: (i) homophones are two words pronounced in the same way but with different spelling; (ii) compound words can be split into their components, as in "outdoors" and "out doors"; (iii) phonetic confusion is the misclassification of one phoneme with a similar one, resulting in the transcription of a different word. The authors also introduce the concept of Skill Squatting Attack, an attack where Alexa opens a (potentially malicious) skill not meant by the user. Lentzsch et al. [12] analyze over 90,000 skills to find out that the Skill Squatting Attack is not being used systematically in the wild, and observe that multiple skills can have the same invocation name, hence, the user could activate a wrong skill.

Security and Privacy in Voice App Skills. The ever-expanding domain of voice app security and privacy has prompted various studies. Both Kumar et al. [11] and Zhang et al. [16] examined threats such as squatting and voice masquerading attacks. Meanwhile, Cheng et al. [6] and Wang et al. [15] assessed the integrity of the skill certification process, uncovering potential loopholes like post-certification code modifications. A notable extension to the voice masquerading attack called the "Alexa versus Alexa" attack, was presented by Esposito [9]. Furthermore, privacy concerns have also received considerable attention. Jide et al. [8] conducted a longitudinal study measuring privacy practices over three years. Other researchers, including Lentzsch et al. [13] examined the comprehensiveness of skills' privacy policies.

7   Discussion

In this study, we explored a specific attack vector targeting Alexa's vehicle-related skills. As we reflect on our findings, it is imperative to address the boundaries of our research and highlight avenues for upcoming investigations.

Scope of Vehicle Skills Tested: Our inquiry predominantly centered around the third-party DroneMobile skill, chosen primarily due to the accessibility it offered concerning vehicle availability. This selection inadvertently excluded car skills from other Original Equipment Manufacturers (OEMs), thereby not fully encompassing the testing potential of Alexa's official car API. We advocate for subsequent studies to branch out and scrutinize skills from diverse OEMs such as Toyota and Land Rover. Such a direction will offer a holistic view of command hijacking threats, not just confined to third-party offerings.

Restrictions on Third-Party Skills: The prevailing market landscape demands device and account linkages for most third-party skills. This stipulation hampered our ability to assess customized third-party offerings exhaustively. We are motivated, in our future endeavors, to delve into any potential conflicts or command hijacking scenarios arising from interactions among diverse third-party skills.

Limitations in Backend Manipulation: The nature of connected car skills mandates rigorous developer verification. This precondition constrained our liberties with backend code manipulation, inevitably capping the range of exploratory actions. An ideal workaround would be to procure access to a developer account specializing in car skills. Such access would empower us to develop and publish bespoke testing skills using Alexa's official API, granting us unrestrained oversight on backend code dynamics.

8   Conclusion

This paper focuses on the research objective of identifying potential vulnerabilities in the Alexa connected vehicle skills. Our investigation has led to the discovery of a novel vulnerability within the intent-matching process of Alexa. This vulnerability can be exploited to develop a new attack that enables the hijacking of Alexa's built-in voice commands, thereby triggering malicious Alexa-connected vehicle skills. In our evaluation, we have provided evidence of the attack's effectiveness by successfully hijacking frequently utilized commands found in commercially connected vehicle skills.

Acknowledgment. This material is based upon work supported in part by the National Science Foundation (NSF) under Grant No. 2239605, 2129164, 2228617, 2120369, 2226339, and 2037798.

References

1. Alexa: The scalable neural architecture behind alexa's ability to select skills. https://www.amazon.science/blog/the-scalable-neural-architecture-behind-alexas-ability-to-select-skills/
2. Amazon: Authorization controller interface. https://developer.amazon.com/en-US/docs/alexa/automotive/alexa-authorizationcontroller.html/
3. Amazon: Connected car skills market. https://www.amazon.com/s?k=vehicle&i=alexa-skills/
4. Amazon: Connected vehicle overview. https://developer.amazon.com/en-US/docs/alexa/automotive/connected-vehicle-overview.html/

W. Ding et al.

5. Amazon: Connected vehicle skills for alexa. https://developer.amazon.com/en-US/docs/alexa/automotive/connected-vehicle-overview.html/
6. Cheng, L., Wilson, C., Liao, S., Young, J., Dong, D., Hu, H.: Dangerous skills got certified: Measuring the trustworthiness of skill certification in voice personal assistant platforms. In: ACM SIGSAC Conference on Computer and Communications Security (CCS) (2020)
7. Compustar: Cs4900-s remote start. https://www.compustar.com/bundles/cs4900-s/
8. Edu, J., Ferrer-Aran, X., Such, J., Suarez-Tangil, G.: Measuring alexa skill privacy practices across three years. In: Proceedings of the ACM Web Conference (WWW), p. 670–680 (2022)
9. Esposito, S., Sgandurra, D., Bella, G.: Alexa versus alexa: controlling smart speakers by self-issuing voice commands. arXiv preprint arXiv:2202.08619 (2022)
10. Kim, Y.B., Kim, D., Kumar, A., Sarikaya, R.: Efficient large-scale neural domain classification with personalized attention. In: Proceedings of the 56th Annual Meeting of the Association for Computational Linguistics (Volume 1: Long Papers), pp. 2214–2224 (2018)
11. Kumar, D., Paccagnella, R., Murley, P., Hennenfent, E., Mason, J., Bates, A., Bailey, M.: Skill Squatting Attacks on Amazon Alexa. In: 27th USENIX Security Symposium (USENIX Security). pp. 33–47 (2018)
12. Lentzsch, C., Shah, S.J., Andow, B., Degeling, M., Das, A., Enck, W.: Hey Alexa, is this skill safe? taking a closer look at the Alexa skill ecosystem. In: Proceedings of the 28th ISOC Annual Network and Distributed Systems Symposium (NDSS) (2021)
13. Lentzsch, C., Shah, S.J., Andow, B., Degeling, M., Das, A., Enck, W.: Hey Alexa, is this skill safe? taking a closer look at the Alexa skill ecosystem. In: 28th Annual Network and Distributed System Security Symposium, NDSS (2021)
14. Seminatore, M.: Alexa tesla. https://github.com/mseminatore/alexa-tesla/
15. Wang, D., Chen, K., Wang, W.: Demystifying the vetting process of voice-controlled skills on markets. Proceedings of the ACM on Interactive, Mobile, Wearable and Ubiquitous Technologies 5(3), 1–28 (2021)
16. Zhang, N., Mi, X., Feng, X., Wang, X., Tian, Y., Qian, F.: Dangerous skills: understanding and mitigating security risks of voice-controlled third-party functions on virtual personal assistant systems. In: 2019 IEEE Symposium on Security and Privacy (SP), pp. 1381–1396 (2019). https://doi.org/10.1109/SP.2019.00016
17. Zhang, Y., Xu, L., Mendoza, A., Yang, G., Chinprutthiwong, P., Gu, G.: Life after speech recognition: fuzzing semantic misinterpretation for voice assistant applications. In: Network and Distributed System Security Symposium (NDSS) (2019)

Enabling Real-Time Restoration of Compromised ECU Firmware in Connected and Autonomous Vehicles

Josh Dafoe, Harsh Singh, Niusen Chen, and Bo Chen[✉]

Department of Computer Science, Michigan Technological University, Michigan, USA

bchen@mtu.edu

Abstract. With increasing development of connected and autonomous vehicles, the risk of cyber threats on them is also increasing. Compared to traditional computer systems, a CAV attack is more critical, as it does not only threaten confidential data or system access, but may endanger the lives of drivers and passengers. To control a vehicle, the attacker may inject malicious control messages into the vehicle's controller area network. To make this attack persistent, the most reliable method is to inject malicious code into an electronic control unit's firmware. This allows the attacker to inject CAN messages and exhibit significant control over the vehicle, posing a safety threat to anyone in proximity.

In this work, we have designed a defensive framework which allows restoring compromised ECU firmware in real time. Our framework combines existing intrusion detection methods with a firmware recovery mechanism using trusted hardware components equipped in ECUs. Especially, the firmware restoration utilizes the existing FTL in the flash storage device. This process is highly efficient by minimizing the necessary restored information. Further, the recovery is managed via a trusted application running in TrustZone secure world. Both the FTL and Trust-Zone are secure when the ECU firmware is compromised. Steganography is used to hide communications during recovery. We have implemented and evaluated our prototype implementation in a testbed simulating the real-world in-vehicle scenario.

Keywords: Connected and autonomous vehicles · ECU · CAN bus · flash translation layer · TrustZone · Steganography

1 Introduction

With rapid growth of automotive industries, both automakers and associated government agencies are taking initiatives to support the development and deployment of connected and autonomous vehicles (CAVs). This includes efforts

Y. Chen et al. (Eds.): SmartSP 2023, LNICST 552, pp. 15–33, 2024.
https://doi.org/10.1007/978-3-031-51630-6_2

to improve CAV efficiency and implement public road infrastructures to support V2V (vehicle-to-vehicle) and V2I (vehicle-to-infrastructure) communications. As this technology continues to develop, CAVs have increasing communication pathways in order to make informed decisions in real time. In addition to V2V and V2I developments, autonomous vehicles are increasingly equipped with many sensors, providing input which will be processed in real time. Especially, increased internal communication is observed, with over 70 electronic control units (ECU) communicating via a in-vehicle network. Recently, in order to manage the numerous ECUs, over-the-air updates have been utilized [13,30]. While these developments allow significant improvement in vehicle functions, they have led to increased security vulnerabilities [17,41,47]. As a result, various attacks on the vehicle systems have been performed by both researchers and real-world hackers. For example, from 2010 to 2018, there were 170 reported automotive attacks, with 60 of these happening in 2018. Further, a plurality of attacks were remote which do not require physical access [39].

One typical attack is performed by accessing the internal network of the vehicle, gaining control over it. Adopted broadly, the controller area network (CAN) provides internal communications among the in-vehicle computer systems. If any malicious entity gains access to the CAN, many in-vehicle operations become vulnerable to manipulation. A few methods have been identified by which the aforementioned access may be gained. First, the attacker hacks remotely into the infotainment system or telematics unit, which manage existing network communication with the outside world. They will then attempt to escalate privileges, and enable the injection of malicious messages which control the victim vehicle. This attack method was demonstrated in 2016 when the Keen Security Lab hacked a Tesla model S [36]. Second, the attacker gains access to the CAN bus physically or remotely via compromising the existing mechanisms for CAN access provided by the on board diagnostics (OBD-II) port. The physical attack occurs through connecting a device directly to the OBD-II port, such as a laptop. However, this is often difficult, as during vehicle usage, the attacker cannot be present, and gaining the initial physical access is challenging. The remote attack through an OBD-II port is performed by compromising an OBD-II dongle that the car owner or mechanic connects [16,31,44]. Also, there are now mobile apps which remotely provide diagnostic services by accessing the CAN [24]. In all the aforementioned attack scenarios, the CAN access is likely limited in time, as the hacker cannot be physically present during the vehicle operation, or the persistence of a remote connection is not guaranteed. Consequently, both the remote and physical attackers would prefer to establish a persistent presence within the CAN. To gain this persistent presence, a best choice for the attacker is to inject malicious code into the internal firmware of ECUs. This occurs via exploiting existing vulnerabilities during firmware updates (such as the over-the-air updates [37]), or ECU programming via the OBD-II port [14]. This work therefore focuses on recovering the ECU firmware which has been compromised by such code injection attacks.

To defend against the ECU code injection attacks, a taxonomy with four categories of CAV defense was established [41], including a passive defense and an active defense. The passive defense framework is focused on detecting and isolating CAV malware attacks, and some additional research has been performed in this area. Specifically, to detect the presence of malicious CAN activity, many intrusion detection systems with high success rates have been evaluated [25–27]. These detection methods observe messages on the CAN bus and establish a model for normal behaviors. Using this model, any malicious deviations are detected. Further, the specific ECU that sent the malicious messages can be detected via analysis of signal characteristics [21,35,46]. In [32], it is observed that relying solely on intrusion detection mechanisms results in a delayed response, where a security update eventually may repair the attack, which is insufficient in the CAV scenario. This is because in this scenario, each moment an attack is active there is more threat posed to both the driver and everyone around them. Since the real-time attack response is so essential, [32] further proposes incorporating an active response by sending detection notifications to the infected ECUs, which will repair via switching to a safe mode and rebooting. However, the specific mechanisms for enabling the restoration of the ECU firmware are still missing. This work thus aims to bridge this gap. Our key observation is that through leveraging trusted hardware components equipped with the ECU, it is possible to enable a real-time restoration of the ECU firmware, which has been compromised by the code injection attacks.

Typically, in-vehicle computers (e.g., ECUs) are equipped with low power processors, including ARM Cortex-A or Cortex-M [1–4] CPUs which are broadly equipped with TrustZone capabilities [10,11]. Further, these same computers may use flash memory as external storage [5,8], which is typically managed by the flash translation layer (FTL). The TrustZone is a hardware-level security feature provided by the processor, which can enable the establishment of a trusted execution environment (TEE) isolated from the normal insecure execution environment. In other words, even if the ECU OS is compromised, the execution running in the TrustZone secure world remains uncompromised. The FTL is a piece of trusted flash memory firmware encapsulated inside a flash storage device (e.g., an SSD drive or a microSD card). It stays between the OS and the flash memory hardware, transparently managing the unique hardware nature of flash memory and exposing externally a block access interface. Therefore, the FTL can also remain secure even if the ECU OS is compromised.

Combining the intrusion detection mechanism with the trusted hardware components, we have established a framework which can efficiently restore the ECU firmware to the version right before the code injection attack (note that we refer to this as the "good prior state" or "good firmware version" throughout the paper). After the compromised ECU is detected by a trusted detection module (i.e., a detector), a notification message will be sent by the detector via the CAN. The notification message will arrive at the compromised ECU and be passed to the trusted application running in the TrustZone secure world via the ECU OS. The trusted application will then collaborate with the trusted FTL to restore the firmware in real time. Our key insight is that whenever a firmware update occurs (e.g., the code injection attack is performed), the FTL, having ultimate control over the underlying storage hardware, can naturally retain an old version of the firmware, due to the out-of-place update feature present in modern flash storage devices. In this way, the FTL can immediately revert to the old good firmware version after the attack. Such a reversion can happen very efficiently as only a small amount of mapping data needs to be restored in the FTL, perfectly meeting the real-time requirement. In addition, to prevent the compromised ECU OS from being aware of the restoration process, all the communications (between the detector and the trusted application, between the trusted application and the FTL) are protected via steganography. In this way, all the communications[1] among them can go through the untrusted ECU OS with the actual purpose being hidden and, the compromised ECU OS typically will not interrupt a seemingly normal process.

2 Background

2.1 Control Area Network

The controller area network (CAN) is a protocol for communication between many nodes connected via two wires where each message is broadcast to all other connected nodes (Fig. 1). Through using this protocol, vehicles are able to greatly reduce the wiring complexity and enable a variable internal network topology. When a node sends a message on the CAN bus, the frame does not include any sender information, but contains a message identifier which describes its type and determines its priority. Based on this identifier, nodes connected to the CAN bus filter out irrelevant messages and accepts those with relevant identifiers. Additionally, each CAN message contains up to 8 bytes of relevant data and commands. The CAN bus is fully accessible, allowing devices or applications to be connected via the on board diagnostics port (OBD-II). In vehicles, CAN is the mechanism for sensors to send data to the main advanced driver assistance systems (ADAS) computer, and for control signals to be sent from the ADAS computer, brake and gas pedals, steering wheel, ignition, etc., to the various ECUs associated with the control operations. Additionally, ECU firmware updates are ultimately sent directly through the CAN bus. Our observation is that when a CAN message is accepted by an ECU, the associated data will be quickly processed [6].

2.2 Flash Memory

Flash memory is broadly used as the external storage device for low-power embedded systems like ECUs [5,8]. This is due to its high throughput, which is necessary in the vehicle scenario, requiring real-time I/O capabilities. Flash

---

[1] Note that the detector should avoid directly communicating with the FTL via the untrusted ECU OS, which is unusual and hence suspicious.

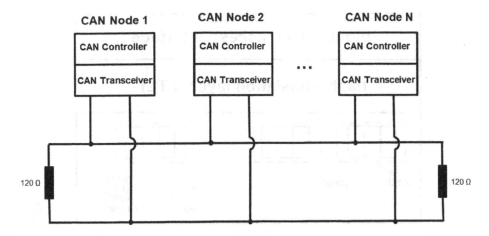

Fig. 1. The topology of a CAN network.

memory (Fig. 2) is organized into a collection of blocks, with each block consisting of smaller pages. However, unique physical characteristics result in differing behavior from hard drive disks (HDD). First, the read/write granularity of flash storage is a page, while the erasure operates on full blocks. Second, each program erase cycle performed on a given block wears down the associated hardware, until a threshold is met and it is considered unreliable and unusable. Due to these special characteristics, in-place updates are expensive. This is because when the data in a single page should be updated, the entire encompassing block must be erased, resulting in further wear. Therefore, an out-of-place update strategy is preferred in which updates are performed by writing the data to a new physical location and marking the old data as invalid. It is also essential to spread program erase cycles throughout the entire storage medium in order to prevent quick wear in any location, so wear-leveling is implemented which handles this. When blocks are invalidated, they are eventually sent to the garbage collector to be erased. Unlike traditional HDDs, the out-of-place update strategy results in different physical locations for the same logical address over time. This is managed by maintaining mappings between physical and logical locations which usually change after each invalidation. All of these firmware components together make up the flash translation layer (FTL), which provides a block access interface externally to the OS. Additionally, the FTL is isolated from the firmware (OS) of its associated ECU by the storage hardware. This isolation provides a guarantee that any computation performed in the FTL will not be compromised even when the ECU firmware is compromised.

2.3   ARM TrustZone

Many ARM processors, such as Cortex-A and Cortex-M CPUs used within automotive ECUs are ARM TrustZone enabled [1–4]. TrustZone establishes a trusted execution environment within a untrusted host. The key idea is to run both

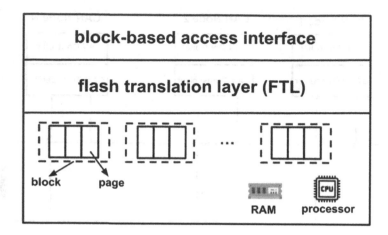

Fig. 2. The architecture of a flash-based block device.

secure (i.e., secure world) and non-secure (i.e., normal world) execution environments on a single processor. The secure world is used to run critical applications with sensitive data, while the normal world can run non-sensitive applications. The two modes are separated by isolating the CPU states and associated memory regions. The architecture of ARM TrustZone is shown in Fig. 3. The communication and interaction between the secure world and the normal world is conducted by secure monitor call (SMC). SMC works as a gateway to ensure invocation of functions and services offered by the secure monitor or secure kernel within the secure world. A salient advantage of TrustZone is that it comes together with the embedded processor and, this hardware-level security feature can be simply utilized without bringing in extra hardware.

2.4 Steganography

Steganography is a mechanism by which to hide some secret message inside of normal data/communications. The secret message is embedded obscurely into original data or messages, such that it goes unnoticed. Different from encryption, this is intended to conceal the fact that a secret message is being sent at all.

3 System and Adversarial Model

3.1 System Model

We consider a connected vehicle with multiple ECUs communicating via the CAN protocol. The ECU is assumed to be equipped with a NAND flash storage device (e.g., an eMMC, a microSD, etc.) on which the ECU firmware is stored. The flash storage device is managed by an FTL, which provides a read/write interface to the ECU OS. The FTL is run on hardware isolated from the OS, so

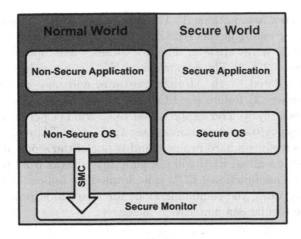

Fig. 3. The architecture of ARM TrustZone.

the computation performed by it is assumed to be secure. Further, each ECU is assumed to be equipped with an ARM processor (Cortex-A or Cortex-M) with TrustZone enabled. Using TrustZone, a trusted world is running in the ECU, on which trusted computation can occur. The trusted world, running trusted applications (TAs) can communicate with untrusted client applications (CAs) running in the untrusted OS (i.e., the potentially compromised ECU firmware). The CAs can perform bidirectional communication with the TA, FTL, and CAN bus. We assume the existence of a trusted in-vehicle computer (IDet) connected to the CAN bus, which performs intrusion detection and signal analysis to detect and localize adversarial ECUs. Note that IDet can communicate directly with the CA via the CAN bus. IDet could be the main ADAS computer or an ECU dedicated to intrusion detection. Our focus in this work is not on malware detection. Therefore, we assume this trusted entity has successfully detected the adversarial ECU [21, 25–27, 35, 46] and we work with the TrustZone and the FTL in the compromised ECU to restore its firmware to a good prior state.

3.2 Adversarial Model

We consider an adversary which can compromise the firmware of an ECU, i.e., by injecting malicious code into the ECU OS. This can be done in a few ways, including remote or physical access to CAN via the OBD-II port, or manipulation of other existing firmware update mechanisms including over-the-air updates. Since the ECU firmware itself is compromised, any detection and recovery mechanisms running in the ECU OS can be averted. This is equivalent to a piece of OS-level malware, which can control the OS of the victim ECU. However, this malware is detectable via intrusion detection of the vehicle, as it must behave maliciously in order to take control of the vehicle, e.g., sending a lot of spoofed CAN messages.

We rely on a few assumptions: 1) The compromised ECU is not able to compromise the TAs running in the TrustZone secure world, which is protected by the processor at the hardware level. This is a common assumption for TrustZone-based applications [23]. 2) The compromised ECU is not able to hack into the FTL, which is isolated by the storage hardware and only presents a limited read/write interface. 3) Before the ECU is compromised, its firmware (OS) is assumed to be healthy. 4) The compromised ECU will not perform DoS attacks, e.g., blocking *regular* communication among CAN, CA, TA, and FTL. Mitigating DoS attacks itself is a hard problem and is out of the scope of this work. In addition, the compromised ECU will not gain any benefits from performing the DoS attacks, as a nonfunctional ECU is an immediate indication of being compromised. In our work, the communications for restoration process are hidden stealthily in the regular communication messages.

4   Design

4.1   Design Overview

Our design consists of four major components (Fig. 4): IDet, CA, TA, and FTL. The IDet (intrusion detector) is running on top of trusted firmware in a secure node, which can communicate with the victim ECU via CAN network. In the victim ECU, there are three components, the CA (client application), the TA (trusted application), and the FTL. The CA is running on top of the untrusted firmware which may be compromised. The TA and the FTL are isolated from the CA by TrustZone hardware and the storage hardware respectively, hence are trusted. We collaborate the aforementioned components in order to restore the compromised ECU firmware to a good prior state after it is compromised.

Our first idea is that the FTL has an ultimate control over the underlying storage hardware and, the previous version of firmware may be maintained and restored. Especially, due to the out-of-place update strategy (Sect. 2.2) in the flash storage, the old version of firmware can be naturally retained in the flash memory blocks, though they will be invalidated when an adversarial update occurs. Since GC eventually erases these invalid blocks, it must be disabled for the old firmware data. Additionally, since the FTL does not know where the firmware is stored, it can be notified before any update occurs, because the firmware is trusted at this point (Sect. 3.2). During recovery, an additional challenge is to find the maintained blocks associated with the good firmware version. These locations can be retrieved by using the mappings associated with the old firmware version, which can be backed up by the FTL.

Our second idea is to securely manage the FTL to restore the ECU firmware even if the entire ECU OS is untrusted. In a vehicle environment, the compromised ECU is able to be identified by another entity (i.e., IDet) outside this ECU in the same vehicle. The IDet needs to inform the FTL to launch the restoration process, but such sensitive messages typically need to go through the CA running on the compromised ECU OS, which will deliver the messages to the

FTL. The direct communication between the IDet and the FTL is very abnormal and the compromised OS will be alerted. Having observed that the data received from the CAN may be processed by the TA running in the TrustZone secure world, and the TA may perform writes to the storage device through the CA [29], our solution is to use the TA as a liaison to forward sensitive messages between the IDet and the FTL. In addition, as the sensitive messages need to go through the untrusted OS, they need to be protected in a plausible manner. Steganography is therefore leveraged to hide the sensitive messages within the regular communications.

Our third idea is to enable the restoration of the ECU firmware when the malware is still present. This is due to the fact that it would be hard for the vehicle user to block the ECU malware once being detected. Upon restoration, the FTL will block all the write requests from the upper layer, and this blocking operation will be canceled once the good firmware has been restored on the external storage and the malware has been removed from the memory.

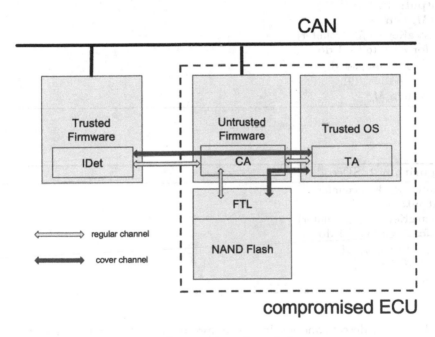

Fig. 4. An overview of our design.

4.2 Design Details

4.2.1 Cover Communications via Steganography.
We define a steganographic message $M_s \in \{0,1\}^k$ to be the message produced when a secret message $\beta \in \{0,1\}^l$ is embedded within a regular cover message $\alpha \in \{0,1\}^k$, where l and

k are both positive integers and $l < k$. Along with the steganographic message are the steganographic algorithms associated with generating and decoding it.

To define the steganographic algorithms used in our design, we first introduce the pseudo random permutation π and pseudo random function f, defined as follows (where s is the length of a shared key):

$$\pi : \{0,1\}^s \times \{0,1\}^{\log_2 k} \to \{0,1\}^{\log_2 k}$$
$$f : \{0,1\}^s \times \{0,1\}^* \to \{0,1\}^s$$

Our steganographic algorithms used during the encoding and decoding processes are defined in Algorithm 1 (SEncode) and 2 (SDecode) respectively. Note that by $f_k(x)$ (or $\pi_k(x)$) we mean applying f (or π) over x using key k.

Algorithm 1. SEncode

Input: β, α, key, counter
Output: M_s
1: $M_s \leftarrow \alpha$
2: stegKey $\leftarrow f_{\text{key}}(\text{counter})$
3: **for** $i = 0$ to $l - 1$ **do**
4:     $j \leftarrow \pi_{\text{stegKey}}(i)$
5:     $M_s[j] \leftarrow \beta[i]$
6: **return** M_s

Algorithm 2. SDecode

Input: M_s, key, counter
Output: β
1: stegKey $\leftarrow f_{\text{key}}(\text{counter})$
2: **for** $i = 0$ to $l - 1$ **do**
3:     $j \leftarrow \pi_{\text{stegKey}}(i)$
4:     $\beta[i] \leftarrow M_s[j]$
5: **return** β

After IDet detects and localizes a compromised ECU, it will send a steganographic message M_{s0} indicating this detection result to the ECU. Since this message will immediately be forwarded to the TA, a key and counter shared between them are used as input to SEncode (Algorithm 1). Further, β is taken to be some secret message, agreed by IDet, TA, and FTL to indicate a malware detection, and α can be any cover message of length k. Unique to the vehicle scenario is that the message is being transmitted over CAN, which has an 8 byte data section that imposes security limitations if using a single CAN frame. Due to this, IDet spreads the k bit message produced by SEncode over $\lceil k \div 64 \rceil$ CAN messages.

After the CA forwards this message (i.e., a collection of $\lceil k \div 64 \rceil$ CAN messages) to the TA, β is extracted using SDecode (Algorithm 2) and is checked against its expected value. If they are identical, a new steganographic message M_{s1} is generated from β with a new α, along with a unique key and counter shared between the TA and the FTL. The TA will send M_{s1} to CA, indicating that it should be written to the FTL. Upon receiving M_{s1}, the FTL will use SDecode to extract β, and check this against its expected value. If they are identical, then firmware restoration will be launched by the FTL. To avoid any replay attacks, both the counter shared between the IDet and the TA, and the counter shared between the TA and the FTL, should be increased by one after each successful restoration.

4.2.2 Firmware Restoration. After receiving the detection notification in the FTL, the old firmware should be restored quickly so that normal operations can resume. To ensure that this is possible, there are two challenges. First, the old firmware must still be present in a recoverable manner on the storage device. Second, the old firmware should be restored quickly to the correct location.

To address the *first* challenge, we exploit the out-of-place updates feature of NAND flash memory. Due to out-of-place updates, during a firmware update the new firmware is written to a different physical location which results in persistence of the old firmware. Normally, when the new data are written to a new physical location, the old blocks are marked as invalid, the mapping from logical address to physical location is updated, and garbage collection (GC) will eventually delete the data. To ensure that the firmware is both maintained and recoverable, we can 1) save the old mappings (from logical to physical location) before an update occurs, and 2) block GC for the blocks associated to relevant saved mappings.

For 1), the FTL reserves a special area for a back up mapping table which stores the saved mappings. Since the ECU firmware is assumed to be trusted prior to the firmware update, a command is sent to a reserved command area by the CA. This command tells the FTL to back up the current mapping tables to the special reserved area. By saving these mappings prior to the update, references to the physical location of the old firmware are maintained. For 2), the data at these physical locations should not be erased. To prevent this, GC is disabled for all blocks invalidated during the firmware update. However, a problem arises when there are multiple firmware updates, as many data blocks will be maintained, but only the mapping tables associated with the last update are preserved. For this reason, GC should be re-enabled on the previously maintained blocks before each firmware update.

To address the *second* challenge, the saved blocks need to be restored. Since the firmware will always boot from the same location, the good firmware should be reverted to this location. To achieve this, the mappings in the reserved area which reference the prior firmware blocks can be restored. Due to this restoration, the same address will now point to the old firmware blocks rather than the malicious firmware. When booting, the ECU will read from the same logical

Fig. 5. Our vehicle testbed.

location as before, but it will point to the physical location of the firmware prior to the adversarial code injection.

4.2.3 Malware Removal. After returning the ECU firmware to the good version prior to the code injection attack, the malware may still be running on the CPU and contained in the ECU memory. A problem associated with this is that the malware can again modify the ECU firmware being restored. Different from the scenario of a real-world computer/mobile device [19], it is hard for the user to block/remove the malware from the victim ECU before the firmware restoration in the FTL after having detecting it. To account for this, once the firmware restoration starts in the FTL, any writes on the FTL should be frozen until the malware has been removed from the memory. To remove the malware from the memory, we can reboot [7] the ECU immediately to clear the memory after its firmware is restored and, after the reboot, the FTL can be notified to cancel the freezing operation.

5   Implementation and Evaluation

To construct the testbed (Fig. 5) with all the necessary components for our implementation, we use two different electronic development boards: 1) Raspberry Pi 3B+ [9] (With 1.4GHz 64-bit quad-core ARM Cortex-A53 CPU, and 1GB LPDDR2 SDRAM) with a RS485 CAN HAT, and 2) a high speed USB header development prototype board LPC-H3131 [33] (with ARM9 32-bit ARM926EJ-S, 180Mhz, 32MB of SDRAM, and 512MB NAND flash). One Raspberry Pi (RPi1) is used as IDet, to generate a detection notification, embed it in regular communication via steganography, and send this over the CAN network. The second Raspberry Pi (RPi2) simply receives the CAN messages and forwards

them via a forwarding socket to the third Raspberry Pi[2]. The third raspberry
pi (RPi3) acts as the infected ECU and the LPC-H3131 is attached to it via a
USB2.0 interface. A client application has been developed to receive CAN messages and forward them to the TA (in RPi3). To develop the TA, we have ported
OP-TEE (Open Portable Trusted Execution Environment) [38] to RPi3. In the
TA, the forwarded message from the CA is received and the secret notification
is extracted. If it is a detection notification, a new message is generated with
the command embedded, which is sent back to the CA for writing to the FTL.
Additionally, we have ported [40] and modified an open source NAND flash manager OpenNFM [22] to the LPC-H3131. Our OpenNFM modifications consist of
backing up the mapping table to a reserved area, reserving a special command
area, decoding the embedded command, and rolling back the previous mappings
to restore the original data. We choose n (number of CAN messages) to be 16,
so for the 64 bit data field of CAN messages, we embed a $l = 64$ bit notification in a 1024 bit message. Consequently, for both embedding and extracting
the notification in IDet, TA, and FTL, we employ the feistel network cipher,
a pseudo-random permutation, with AES as the round function to produce a
$\log_2 1024 = 10$ bit output.

Evaluating Recovery and Communication. To evaluate the time for processing the hidden communications and recovering the ECU firmware to before
an attack, we timed three phases: 1) time in IDet (encoding the detection notification), 2) time in TA (extracting the notification in TA, embedding the command in a new message, and sending the new message to FTL via a write request
to the CA), and 3) time in FTL (extracting the notification in the FTL and subsequently performing recovery). The time of each phase has been measured 20
times and the results are summarized in Table 1. These results demonstrate that
after malicious CAN messages are detected via intrusion detection and signal
analysis mechanisms, a hidden detection notification can quickly be transmitted to the compromised ECU and the firmware can be restored to the prior
version. Note that the majority of time spent in the FTL is for recovery, with
around .01 s for decoding the notification. The time in TA is significantly more
than expected. This is likely due to the overhead from computation performed
(extracting the notification, generating a new message and embedding the command, sending this back to the CA), context switching overhead, and the limited
physical resources available in the secure world.

Table 1. Average time (measured in seconds) in IDet, TA, and FTL.

IDet Time (s)	TA Time (s)	FTL Time (s)
.0012988	7.1065939	1.4264553

---

[2] This is necessary due to unresolved compatibility issues between the CAN drivers
and specific OP-TEE implementation.

Throughput Evaluation To evaluate the impact of our solution on normal operations with the flash storage device, we performed a throughput comparison between the unmodified OpenNFM and our modified version. For these benchmarks, we used the fio benchmarking tool [12] and performed random write (RW), random read (RR), sequential write (SW), and sequential read (SR) benchmarks. Our results from these tests are summarized in Fig. 6.

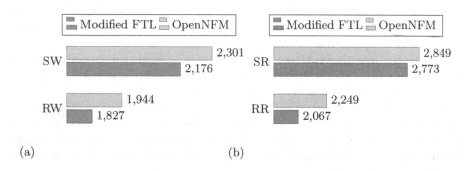

Fig. 6. Throughput comparison (KB/s).

To fully evaluate the impact of the observed differences, we perform a statistical analysis of the results. The values reflected in Fig. 6 are the mean of 30 benchmarks for each type. Evaluations of the differences in throughput are based on the mean values. Subsequently, a T-Test was performed to determine the statistical significance of the observed results for RW (random write), RR (random read), SW (sequential write), and SR (sequential read). The p values were $< .00001$ for all tests, which is less than the .05 level, indicating a statistically significant difference in throughput. These throughput results demonstrate that the implementation of our solution results in a small performance cost on normal read/write operations. Evaluating the extent of these differences, the average difference in throughput is 5.7%.

6   Related Work

6.1   Data Recovery in Embedded Systems

Chen et al. developed FFRecovery [18], an fine-grained data recovery system specifically designed for flash memory devices. FFRecovery employs file system forensics techniques to restore metadata, while utilizing the out-of-place update feature in the flash translation layer to extract raw data. Xie et al. proposed MobiDR, a data recovery framework for mobile devices [45]. MobiDR can recover user data to the corruption point to defend against OS-level malware by utilizing the cloud server's version control capability and the hardware features of the local device. Guan et al. introduced Bolt, a system aimed at restoring the operating system without requiring a reboot. Bolt makes use of ARM TrustZone to backup the memory snapshot and utilizes a customized firmware to save the snapshot of flash memory. This approach enables efficient recovery of the system to a clean state after it is compromised. Huang et al. presented FlashGuard [28], a firmware-level recovery system against ransomware attacks that provides rapid and efficient data recovery without the need for explicit backups. FlashGuard makes use of the out-of-place update feature in solid state drive to retrieve data encrypted by ransomware. Wang et al. designed TIMESSD [43]. To recover user data that are compromised by the malware, TIMESSD leverages the out-of-place update characteristics in SSD to retain the states of history storage, and the recovery process is achieved via calling rollback functions. Furthermore, Baek et al. [15], Wang et al. [42], and Min et al. [34] have integrated the ransomware detection component into the flash translation layer. This integration can save the space that would otherwise be allocated for backing up invalid data in local storage if ransomware is not detected.

Different from the existing works for data recovery in the embedded systems, our design targets the ECU firmware recovery in a vehicle system which is much more critical than the traditional embedded systems. This type of critical systems has a much higher demand in a few aspects. First, it requires the restoration to be performed in real time, i.e., the extra overhead caused by the design should be minimized. Second, upon restoration, the malware is still present as it is difficult for the user/driver to block the malware in the compromised ECU, i.e., the restoration needs to happen when the malware is still present. Third, the malware detector comes from another entity in the vehicle and the communication between the detector and the FTL need to be adapted to the communication network inside a vehicle. Also, it needs to be protected and avoids being disturbed by the compromised ECU firmware.

6.2 CAN Intrusion and Malicious ECU Defenses

Kwon et al. proposed a mitigation mechanism against CAN intrusion by configuring impacted ECUs and ignoring malicious message IDs [32]. They proposed using intrusion detection to send messages to malicious ECUs and reconfigured them to a good state, but this state provides less functionality, and they did not provide any implementation or mechanism toward providing these guarantees. Han et al. evaluated the use of survival analysis for intrusion detection, with higher than 97% detection accuracy over all tested vehicles [26]. Yang et al. [46] designed a mechanism for detecting spoofing attacks from unrecognized ECUs by authenticating CAN data frame IDs. Using a recurrent neural network (RNN), they authenticated sender identity based on fingerprint signals. Further, Cho and Shin used recursive least squares to construct a baseline of ECU clock behavior for developing an anomaly-based intrusion detection system which identifies the malicious ECU [20].

7 Conclusion

In this work, we have designed a new framework for connected and autonomous vehicles to defend against the ECU code injection attacks, by rolling back the compromised ECU firmware to a good prior state. Our design has taken advantage of various existing hardware features equipped with the ECU to securely manage and efficiently perform the recovery process. We have implemented a prototype for the proposed framework and demonstrated its effectiveness at performing real-time recovery in a simulated in-vehicle testbed.

Acknowledgments. This work was supported by US National Science Foundation under grant number 2225424-CNS, 1928349-CNS, and 2043022-DGE.

References

1. https://www.nxp.com/products/processors-and-microcontrollers/s32-automotive-platform/s32z-and-s32e-real-time-processors:S32Z-E-REAL-TIME-PROCESSORS
2. https://www.xilinx.com/products/boards-and-kits/zcu104.html#information
3. https://www.nxp.com/design/designs/s32g3-vehicle-networking-reference-design:S32G-VNP-RDB3
4. https://www.nxp.com/design/designs/s32k3-automotive-telematics-box-t-box-reference-design-board:S32K3-T-BOX
5. Autonomous vehicle data storage - premio inc. https://premioinc.com/pages/autonomous-vehicle-data-storage
6. Can - Automotive Basics. https://automotivetechis.wordpress.com/2012-06-01-can-basics-faq/
7. Everything you need to know about performing an ECU reset. https://www.way.com/blog/ecu-reset/
8. Memory use in automotive - electronic products. https://www.electronicproducts.com/memory-use-in-automotive/
9. Raspberry pi 3 model b+. https://www.raspberrypi.com/products/raspberry-pi-3-model-b-plus/
10. Trustzone for cortex-a - arm®. https://www.arm.com/technologies/trustzone-for-cortex-a
11. Trustzone for cortex-m - arm®. https://www.arm.com/technologies/trustzone-for-cortex-m
12. fio (2014). http://freecode.com/projects/fio
13. Applying over-the-air updates in safely automotive ECUS (2021). https://www.nxp.com/company/blog/applying-over-the-air-updates-in-safely-automotive-ecus:BL-OTA-IN-AUTO-ECUS
14. ECU programming guide (2021). https://ecutek.zendesk.com/hc/en-gb/articles/207345569-ECU-programming-guide

15. Baek, S., Jung, Y., Mohaisen, A., Lee, S., Nyang, D.: SSD-insider: internal defense of solid-state drive against ransomware with perfect data recovery. In: 2018 IEEE 38th International Conference on Distributed Computing Systems (ICDCS), pp. 875–884. IEEE (2018)

16. Bielawski, R., Gaynier, R., Ma, D., Lauzon, S., Weimerskirch, A.: Cybersecurity of firmware updates. Technical Report DOT HS 812 807, University of Michigan. Transportation Research Institute and University of Michigan, Dearborn and Volkswagen Group of America (Herndon, VA) (October 2020), https://rosap.ntl. bts.gov/view/dot/55729

17. Chattopadhyay, A., Lam, K.Y., Tavva, Y.: Autonomous vehicle: Security by design. IEEE Trans. Intell. Transp. Syst. **22**(11), 7015–7029 (2021). https://doi.org/10. 1109/TITS.2020.3000797

18. Chen, N., Dafoe, J., Chen, B.: Poster: data recovery from ransomware attacks via file system forensics and flash translation layer data extraction. In: Proceedings of the 2022 ACM SIGSAC Conference on Computer and Communications Security, pp. 3335–3337 (2022)

19. Chen, N., Xie, W., Chen, B.: Combating the OS-level malware in mobile devices by leveraging isolation and steganography. In: Zhou, J., et al. (eds.) Applied Cryptography and Network Security Workshops: ACNS 2021. LNCS, vol. 12809, pp. 397–413. Springer, Cham (2021). https://doi.org/10.1007/978-3-030-81645-2_23

20. Cho, K.T., Shin, K.G.: Fingerprinting electronic control units for vehicle intrusion detection. In: 25th USENIX Security Symposium (USENIX Security 16), pp. 911–927. USENIX Association, Austin (2016). https://www.usenix.org/conference/usenixsecurity16/technical-sessions/presentation/cho

21. Choi, W., Jo, H.J., Woo, S., Chun, J.Y., Park, J., Lee, D.H.: Identifying ECUS using inimitable characteristics of signals in controller area networks. IEEE Trans. Veh. Technol. **67**(6), 4757–4770 (2018). https://doi.org/10.1109/TVT.2018. 2810232

22. Code, G.: Opennfm. https://code.google.com/p/opennfm/

23. Guan, L., et al.: Supporting transparent snapshot for bare-metal malware analysis on mobile devices. In: Proceedings of the 33rd Annual Computer Security Applications Conference, pp. 339–349 (2017)

24. Hackenberg, R., Weiss, N., Renner, S., Pozzobon, E.: Extending vehicle attack surface through smart devices (2017)

25. Hamada, Y., Inoue, M., Ueda, H., Miyashita, Y., Hata, Y.: Anomaly-based intrusion detection using the density estimation of reception cycle periods for in-vehicle networks. SAE Int. J. Transport. Cybersecur. Privacy **1** (2018). https://doi.org/10.4271/11-01-01-0003

26. Han, M.L., Kwak, B.I., Kim, H.K.: Anomaly intrusion detection method for vehicular networks based on survival analysis. Veh. Commun. **14**, 52–63 (2018). https://doi.org/10.1016/j.vehcom.2018.09.004

27. Hoppe, T., Kiltz, S., Dittmann, J.: Applying intrusion detection to automotive it-early insights and remaining challenges. J. Inf. Assur. Secur. (JIAS) **4**, 226–235 (2009)

28. Huang, J., Xu, J., Xing, X., Liu, P., Qureshi, M.K.: Flashguard: leveraging intrinsic flash properties to defend against encryption ransomware. In: Proceedings of the 2017 ACM SIGSAC Conference on Computer and Communications Security, pp. 2231–2244 (2017)

29. Köhler, J., Förster, H.: Trusted execution environments in vehicles. ATZelektronik worldwide **11**(5), 36–41 (2016). https://doi.org/10.1007/s38314-016-0074-y

30. Kim, B., Park, S.: ECU software updating scenario using OTA technology through mobile communication network. In: 2018 IEEE 3rd International Conference on Communication and Information Systems (ICCIS), pp. 67–72. IEEE (2018)
31. Klinedinst, D.J., King, C.: On board diagnostics: Risks and vulnerabilities of the connected vehicle. CERT Division, Software Engineering Institute, Carnegie Mellon University, April, White paper (2016)
32. Kwon, H., Lee, S., Choi, J., Chung, B.H.: Mitigation mechanism against in-vehicle network intrusion by reconfiguring ECU and disabling attack packet. In: 2018 International Conference on Information Technology (InCIT), pp. 1–5 (2018). https://doi.org/10.23919/INCIT.2018.8584882
33. Ltd., O.: Lpc-h3131. https://www.olimex.com/Products/ARM/NXP/LPC-H3131/. Accessed 30 June 2023
34. Min, D., et al.: Amoeba: an autonomous backup and recovery SSD for ransomware attack defense. IEEE Comput. Archit. Lett. **17**(2), 245–248 (2018)
35. Murvay, P.S., Groza, B.: Source identification using signal characteristics in controller area networks. IEEE Signal Process. Lett. **21**(4), 395–399 (2014). https://doi.org/10.1109/LSP.2014.2304139
36. News, T.H.: Hackers take Remote Control of Tesla's Brakes and Door locks from 12 Miles Away. https://thehackernews.com/2016/09/hack-tesla-autopilot.html
37. Nie, S., Liu, L., Du, Y., Zhang, W.: Over-the-air: how we remotely compromised the gateway, BCM, and autopilot ECUs of tesla cars. Briefing, Black Hat, vol. 91 (2018)
38. OP-TEE. Op-tee documentation. https://optee.readthedocs.io/en/latest/general/about.html Accessed 30 June 2023
39. Stevebell. A Pivotal Year for Black Hat Cyber Attacks on Connected Cars - TU Automotive (2008). https://www.tu-auto.com/2018-a-pivotal-year-for-black-hat-cyber-attacks-on-connected-cars/
40. Tankasala, D., Chen, N., Chen, B.: A step-by-step guideline for creating a testbed for flash memory research via LPC-h3131 and opennfm (2020)
41. Thing, V.L., Wu, J.: Autonomous vehicle security: a taxonomy of attacks and defences. In: 2016 IEEE International Conference on Internet of Things (iThings) and IEEE Green Computing and Communications (GreenCom) and IEEE Cyber, Physical and Social Computing (CPSCom) and IEEE Smart Data (SmartData), pp. 164–170 (2016). https://doi.org/10.1109/iThings-GreenCom-CPSCom-SmartData.2016.52
42. Wang, P., Jia, S., Chen, B., Xia, L., Liu, P.: Mimosaftl: adding secure and practical ransomware defense strategy to flash translation layer. In: Proceedings of the Ninth ACM Conference on Data and Application Security and Privacy, pp. 327–338 (2019)
43. Wang, X., Yuan, Y., Zhou, Y., Coats, C.C., Huang, J.: Project almanac: a time-traveling solid-state drive. In: Proceedings of the Fourteenth EuroSys Conference 2019, pp. 1–16 (2019)
44. Wen, H., Chen, Q.A., Lin, Z.: Plug-N-Pwned: comprehensive vulnerability analysis of OBD-II dongles as a new Over-the-Air attack surface in automotive IoT. In: 29th USENIX Security Symposium (USENIX Security 20), pp. 949–965. USENIX Association (2020). https://www.usenix.org/conference/usenixsecurity20/presentation/wen
45. Xie, W., Chen, N., Chen, B.: Enabling accurate data recovery for mobile devices against malware attacks. In: 18th EAI International Conference on Security and Privacy in Communication Networks (2022)

46. Yang, Y., Duan, Z., Tehranipoor, M.: Identify a spoofing attack on an in-vehicle can bus based on the deep features of an ECU fingerprint signal. Smart Cities **3**(1), 17–30 (2020). https://doi.org/10.3390/smartcities3010002

47. Zhang, T., Antunes, H., Aggarwal, S.: Defending connected vehicles against malware: challenges and a solution framework. IEEE Internet Things J. **1**(1), 10–21 (2014). https://doi.org/10.1109/JIOT.2014.2302386

mmFingerprint: A New Application Fingerprinting Technique via mmWave Sensing and Its Use in Rowhammer Detection

Sisheng Liang[1]($\boxtimes$), Zhengxiong Li[2], Chenxu Jiang[3], Linke Guo[3], and Zhenkai Zhang[1]

[1] School of Computing, Clemson University, Clemson, UK
{sishenl,zhenkai}@clemson.edu
[2] Department of Computer Science and Engineering, University of Colorado Denver, Denver, USA
zhengxiong.li@ucdenver.edu
[3] Holcombe Department of Electrical and Computer Engineering, Clemson University, Clemson, UK
{chenxuj,linkeg}@clemson.edu

Abstract. Application fingerprinting is a technique broadly utilized in diverse fields such as cybersecurity, network management, and software development. We discover that the mechanical vibrations of cooling fans for both the CPU and power supply unit (PSU) in a system strongly correlate with the computational activities of running applications. In this study, we measure such vibrations with the help of mmWave sensing and design a new application fingerprinting approach named mmFingerprint. We create a prototype of mmFingerprint and demonstrate its effectiveness in distinguishing between various applications. To showcase the use of mmFingerprint in cybersecurity for defensive purposes, we deploy it in a real computer system to detect the execution of reputable Rowhammer attack tools like TRRespass and Blacksmith. We find that the detection can reach a very high accuracy in practical scenarios. Specifically, the accuracy is 89% when exploiting CPU fan vibrations and nearly 100% when leveraging PSU fan vibrations.

Keywords: Application fingerprinting · mmWave sensing · physical side-channel · Rowhammer detection

1 Introduction

Fingerprints are unique attributes that objects possess, and can be used to differentiate one from another despite their similarities [1]. This concept naturally extends into the digital world, where we see its application in the form of application fingerprinting. Generally speaking, application fingerprinting is a process that identifies, detects, and catalogs running applications based on distinctive elements, such as patterns in data usage, computation/network behavior, or specific configurations within the application's code.

© ICST Institute for Computer Sciences, Social Informatics and Telecommunications Engineering 2024
Published by Springer Nature Switzerland AG 2024. All Rights Reserved
Y. Chen et al. (Eds.): SmartSP 2023, LNICST 552, pp. 34–52, 2024.
https://doi.org/10.1007/978-3-031-51630-6_3

In recent years, application fingerprinting techniques have been widely employed in various areas, including cybersecurity, network management, and software development. As representative examples in cybersecurity, not only can these techniques be exploited for compromising user privacy [2,3], but they can also be employed for defensive purposes, such as detecting the use of illicit programs (e.g., those for crypto mining and password cracking) on high-performance computing systems [4,5] and identifying the execution of denial-of-service (DoS) or other malicious software [6,7].

The practice of application fingerprinting leveraging side-channel information has gained considerable popularity. This is because side-channel information, such as power consumption [8], and electromagnetic radiation [3,9–11], are inevitable byproducts of any computation and can be hardly suppressed by external adversaries [12]. More importantly, the information correlates with the ongoing computation activities, making side-channel-based application fingerprinting possible.

In this paper, we propose a novel approach leveraging certain physical side-channel information obtained through mmWave sensing to achieve application fingerprinting that can be used to replace or complement traditional application fingerprinting methods as present in Fig. 1. The foundation of our approach is built on the observation that different applications generate varying computation activities, which modulate the speed of the cooling fan. These modulated cooling fan speeds can reveal the computation activities. Therefore, accurately measuring these speed variations becomes the key. Equipped with advanced range and vibration sensing techniques, mmWave sensing, our method can measure fine speed variations with high precision. By monitoring the vibration patterns incurred by the speed of the cooling fan, our technique employs features engineering and deep neural networks to extract features and then uses a deep learning classifier to distinguish the applications.

Compared to the conventional application fingerprinting methods using network traffic statistics [13,14], our approach has the following advantages: (1) We can indirectly monitor the computational actions of an application through the fan's status. This is particularly beneficial when the application does not generate any network traffic or when some applications alter the characteristics of the network traffic to make it seem legitimate [15]. (2) Our system provides non-intrusive and remote monitoring. It cannot be easily suppressed by external adversaries due to the contact-less fashion. (3) It does not add performance overhead to the target computing system.

Alongside the introduction of our new application fingerprinting technique, we also demonstrate its practical use in the field of detecting the execution of malicious programs. Specifically, we show that our fingerprinting technique can accurately identify potential Rowhammer attempts carried out by certain existing tools. We concentrate on this type of threat for two main reasons: the severity of Rowhammer attacks and the prevalent use of established tools in the initial reconnaissance phase.

Firstly, Rowhammer attacks pose substantial and ongoing threats to computer systems, leading to numerous exploitations such as sandbox escaping, privilege escalation [16,17], cryptography subversion [18], denial of service [19–21], and even confidentiality breaches [22]. Although there are many mitigation strategies proposed, including counter-based methods such as [23–25], and Target Row Refresh (TRR) that is implemented in the current off-the-shelf DDR4 DRAMs by major vendors. However, advanced Rowhammer attack techniques such as TRRespass [26] and Blacksmith [27] have circumvented TRR. The effectiveness of counter-based defenses becomes questionable for this new type of many-sided Rowhammer attack.

Secondly, before launching a real Rowhammer attack, an attacker must inspect and scan the system to determine if its memory is susceptible to the Rowhammer effect. It is highly likely that during this reconnaissance phase, the attacker will utilize one or more reputable and effective tools, such as TRRespass [26] and Blacksmith [27], for such a purpose. These tools are known for their efficiency in hammering standard DDR4 DRAM modules, even those under the protection of TRR, aiding the attacker in swiftly identifying exploitable bits.

We evaluate mmFingerprint using data gathered from a CPU cooling fan and a PSU cooling fan, each subjected to ten different applications. These include two of the latest and most potent Rowhammer attack tools as well as harmless applications like the SPEC 2006 benchmark, YouTube, and system idle states. mmFingerprint demonstrates robust performance across these applications, achieving accuracy ranging from 0.69 to 1.00 in various scenarios. Notably, it can detect known Rowhammer attacks with near-perfect accuracy. Our findings indicate that the approach we've introduced is a feasible method for detecting Rowhammer attacks when established tools are used during the preliminary reconnaissance phase.

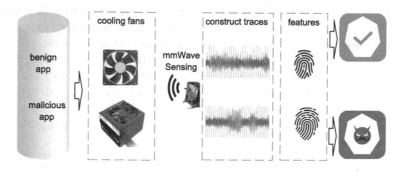

Fig. 1. mmFingerprint is based on monitoring the fan status through mmWave sensing and it can be used to detect if malicious applications are running.

The main contributions of this paper include:

– We introduce an innovative approach to application fingerprinting that capitalizes on side-channel information from cooling fans and mmWave sensing technology. This method identifies applications by picking up the subtle vibration differences on the cooling fan induced by the computation activities. To the best of our knowledge, this is the first time that mmWave sensing has been applied to the context of application fingerprinting.

- We exemplify its defensive application by illustrating how it can detect Rowhammer attacks executed with recognized hammering tools during the reconnaissance process. We are the first to introduce mmWave sensing in the detection of Rowhammer attacks. It provides a new research vision in this area.
- The proposed mmFingerprint can efficiently recognize the most sophisticated Rowhammer attempts with reputable tools during the reconnaissance phase. The accuracy of this method can reach up to 100% percent.

2   Background

2.1   Advanced Configuration and Power Interface

The Advanced Configuration and Power Interface (ACPI) specification is an industry-wide standard that enables sophisticated operating system-directed configuration and power management for both individual devices and whole systems via the motherboard. [28]. It is comprised of both software and hardware elements. Devices and processors can run on different states based on the necessity to maintain a balance among power saving, heat dissipation, and performance. For example, it defines four useful states for a processor: the C0 state, where the CPU is doing useful work; the C1 (Halt) state, a light sleep state where the processor isn't executing instructions; the C2 (Stop-Clock) state, a deeper sleep state where power to the core is shut off; and the C3 (Sleep) state, an even deeper sleep state where the cache's context is lost and power to the cache is shut off. ACPI allows the OS to play a role in the thermal management of the system while maintaining the platform's ability to mandate cooling actions as necessary. It defines two cooling modes, Active and Passive. In the passive cooling mode, OS reduces the power consumption of devices at the cost of system performance to reduce the temperature of the system. While in active cooling, OS increases the power consumption of the system (for example, turning on a fan) to reduce the temperature of the system [28].

The OS active cooling mode needs support from the hardware such as the thermal sensor, cooling fan, and fan speed controller. The cooling fans are important computer components that help dissipate the heat generated by electronic components such as CPU, GPU, and the power electronics in the power supply. Most modern computer systems use temperature-controlled fan speed control mechanisms to regulate CPU and GPU cooling fan speeds. These mechanisms use hardware sensors to monitor CPU temperature and adjust the fan speed accordingly. Usually, the speed is a function of the temperature. This function can be selected from different working modes in the BIOS of some modern motherboards. The speed control approaches described include on-off, linear, and pulse width modulation (PWM) [29].

2.2   MmWave Sensing

The high-resolution frequency-modulated continuous-wave (FMCW) mmWave radar has been widely used in automotive and industrial applications recently due to the low cost [30]. It can be used to detect objects by estimating the range, velocity, and angle [31]. The mmWave radar transmits serial FMCW signals and receives the corresponding reflection signals from surrounding objects. Mixing the transmitted signal and received signal produces an intermediate-frequency (IF) signal, which can be used to estimate range, velocity, and angle. By tracking changes in the estimated range over a specific time step, the variation can be considered the object's vibration. The derivation of vibration is widely used in speech eavesdropping and reconstruction [32–34], vibration monitoring [35].

The estimation of range with coarse resolution can be achieved by applying a range FFT to the IF signal. With a 4GHz bandwidth FMCW mmWave device, the resolution stands at 3.75 cm [33]. This level of resolution suffices for many applications, like detecting objects in automotive settings. However, it falls short for applications that need a higher degree of detail, such as sound reconstruction and subtle vibration tracking, which typically require finer resolution. For these applications, a high-resolution range (e.g. 1 mm or even smaller) can be extracted from the phase value corresponding to the target range.

2.3   Rowhammer Attacks

Rowhammer attacks are a class of security exploits that target a hardware vulnerability in dynamic random-access memory (DRAM). By repeatedly accessing some DRAM rows, an attacker can cause unintended bit flips in neighboring rows by accelerating capacitor charge leakage, potentially leading to unauthorized access or privilege escalation, etc. The execution of a Rowhammer attack involves three phases by the attackers [12].

- Phase 1, the attacker scans the DRAM addresses by repeatedly accessing certain DRAM rows to search for exploitable bit flips. For example, with the addresses mapping information obtained by reverse engineering before the attack, the attacker can explore Rowhammer scanning by accessing two addresses from the same bank but not in the same row. When bit flips are found, the attacker can record the corresponding physical address for later use.
- Phase 2, The attacker redirects the target's sensitive security data to the vulnerable location identified in the first step.
- Phase 3, the attacker flips the bits when the security-critical data is placed at the location where it is flippable according to the second step. Then, the attacker can achieve his design goals such as privilege escalation, cryptography subversion, denial of service, and confidentiality breaching from this step.

Major DRAM vendors have widely adopted the Targeted Row Refresh (TRR) strategy to counteract Rowhammer attacks on the DDR4 memory. When the number of accesses to a particular row surpasses a set threshold, a refresh (or activation) is issued to the neighboring rows. This action recharges these adjacent rows, thereby safeguarding them from being flipped. However, several advanced Rowhammer tools have recently been developed to bypass this TRR mitigation strategy, implemented by leading manufacturers on certain DDR4 DRAMs. Examples of such tools include TRRespass [26] and Blacksmith [27]. These tools are typically employed by attackers during the reconnaissance phase of a Rowhammer attack due to their efficacy. TRRespass utilizes a many-sided hammering technique to trigger bit flips and circumvent the TRR by generating a high volume of accesses to different DRAM rows in the same bank during the refresh window. Meanwhile, Blacksmith optimizes the row access pattern to achieve higher efficiency than TRRespass in triggering bit flips by adjusting the offset and intensity of hammering.

3 mmFingerprint

In this section, we present a robust technique called mmFingerprint, designed for application fingerprinting in systems that incorporate a CPU cooling fan or a power supply fan. These applications impact the CPU temperature or power electronics in the PSU, which subsequently alters the speed of the CPU fan or PSU fan. The mmFingerprint tool is adept at identifying such minor shifts in fan speed. The system can differentiate among various applications by analyzing the vibrations in the CPU cooling fan or power supply unit (PSU) fan, without requiring direct physical interaction. mmFingerprint employs advanced signal processing methods to detect these subtle vibrations.

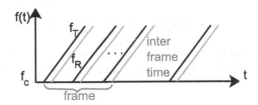

Fig. 2. mmWave FMCW chirps

3.1 Estimating Displacement Using MmWave Technology

mmWave radar adopts the Frequency Modulated Continuous Wave (FMCW) chirps for distance measuring [31]. Estimating the distance between the transmitter and receiver can be achieved by measuring the time delay and phase shift of mmWave signals. Furthermore, mmWave sensing enables the determination of object displacement by analyzing the range difference of the same object over a given time interval. For example, given a sinusoidal FMCW transmit signal represented by

$$x_T(t) = \cos\left(2\pi f_c t + \pi S t^2\right) , \tag{1}$$

where f_c is the start frequency of the chirp and S is the frequency slope of the chirp. The time delay between the transmitter signal and the receiver signal can be represented as

$$\tau = 2d/c , \tag{2}$$

where τ is the time delay; d is the distance from the antenna to the target; c denotes the speed of light. The mixer combines the incoming and outgoing signals to generate the intermediate frequency (IF) signal. After the high-frequency components are eliminated by a low-pass filter, the low-frequency elements remain in the IF signal, which can be represented by

$$x_{IF}(t) = \text{LPF}\left\{x_T(t)x_R(t)\right\} = A\cos\left(2\pi f_{IF}t + \phi_{IF}\right) . \tag{3}$$

The intermediate frequency f_{IF} can be represented by the difference between the transmit signal frequency $f_T(t)$ and receiver signal frequency $f_R(t)$, as shown in

$$f_{IF} = f_T(t) - f_R(t) = S\tau , \tag{4}$$

which can be further represented by the chirp frequency slope S and time delay τ according to the geometric relationship between the intermediate frequency and the frequency slope of the chirp as presented in Fig. 2. The intermediate signal initial phase can be determined from (1) at the time instant τ when the reflected signal just arrives at the antenna, which can be represented as

$$\phi_{IF} = 2\pi f_c \tau + \pi S \tau^2 \approx 2\pi f_c \tau. \tag{5}$$

It can be approximated because f_c is much larger than $S\tau$ [31].

Finally, from (2) and (4) the distance and frequency relation can be represented as

$$d = S\tau = cf_{IF}/(2S). \tag{6}$$

By performing the FFT operation to the intermediate signal (range FFT), the ranges can be obtained according to this equation. However, the range resolution is only 3.75 cm for a 4 GHz continuous bandwidth mmWave radar such as the TI IWR1642BOOST since the range resolution is determined by $c/(2B)$, where B is the chirp bandwidth [36]. This resolution is enough for applications such as distance detection in vehicles. However, it is not effective for applications requiring 1-mm or even better resolution such as voice recovery. Fortunately, we can derive a high-resolution range from phase based on (2) and (5), which can be represented as

$$\phi_{IF} = 2\pi f_c \tau = 4\pi d/\lambda , \tag{7}$$

where λ is the wavelength of mmWave signal at frequency f_c. Differentiating both sides of the Eq. (7) results in $\Delta d = \lambda \Delta \phi_{\mathrm{IF}}/4\pi$, where Δd is the small range displacement for a target during a short time; $\lambda \approx 4mm$ is mmWave signal wavelength for a 77-81Ghz mmWave radar. $\Delta \phi_{\mathrm{IF}}$ is the corresponding phase displacement for the same target. The displacement calculated through phase yields a better range resolution than that derived from range FFT which is 3.75 cm for a 4Ghz bandwidth mmWave radar.

3.2   Locate the Cooling Fan with MmWave Radar

First, mmFingerprint locates the target cooling fan with mmWave sensing. mmFingerprint conducts a range-FFT over each chirp on the gathered Intermediate Frequency (IF) data. Different frequency components represent distinct reflective signals from various objects in the surrounding environment. Identifying the desired frequency bin (range bin) among numerous bins can be challenging. We monitor various range bins across several consecutive frames, as shown in Fig. 3. Each peak represents an object. We identify the correct range bin by locating the right peak and verifying it with a measured distance from a ruler. Second, once the target range bin has been located, mmFingerprint extracts the phase value at the target bin by calculating the phase angle from the complex values at the peak. According to Eq. (7), the phase value is proportional to the target distance.

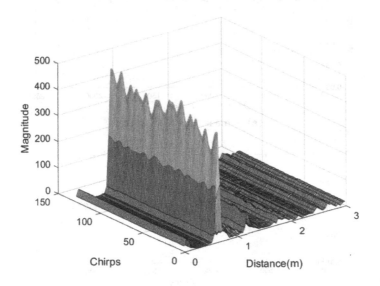

Fig. 3. Range-FFT across many chirps.

3.3   Time Serials Trace Construction

Using the phase data from the designated bin, `mmFingerprint` initially creates the distance-time series traces for a targeted object. It does this by joining together the phase values obtained at the targeted bin from every chirp, over a multitude of continuous frames. Then, range displacement is derived from the distance-time series trace according to $\Delta d_n = d_{n+1} - d_n$, where d_{n+1} is the distance at discrete time $n+1$ and d_n is the distance at discrete time n. Therefore, the range displacement is sampled at the sample rate of the chirp rate.

Removing the Spikes: The mmWave radar produces chirps in frames, in a non-continuous fashion. There is a noticeable surge at the start of each frame due to the first two data points, and these surges significantly exceed other phase values as shown in Fig. 4(a). In order to mitigate the influence of these abnormal data points on the classification process, we replace them with the final data point from the preceding frame. This strategy facilitates a seamless transition from one frame to the next. As shown in Fig. 4(b), the range displacement trace oscillates around zero in a more symmetrical way. The useful side-channel information encoded into the recovered time series trace can be exploited to infer the computing activities.

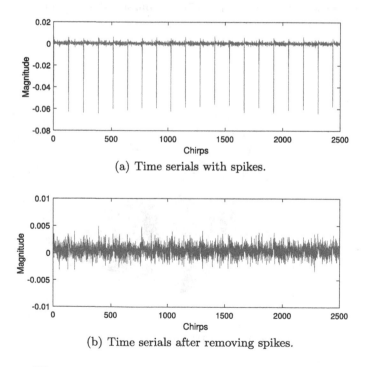

(a) Time serials with spikes.

(b) Time serials after removing spikes.

Fig. 4. Reconstructed displacement time serials traces.

3.4 Fan Responses and Correlations

A time series trace can be constructed from the displacement derived from the phase with the aforementioned method. Applying the Fast Fourier Transform (FFT) to the time domain signal is a common technique used to analyze frequency components and extract features from signals. By converting the signal from the time domain to the frequency domain, we can examine the distribution of frequencies present in the signal and identify specific patterns or characteristics. Figure 6 presents the frequency components of the two time-series traces for different loads. The frequency distributions for these two traces display uniqueness. When the CPU executes different applications, the computational tasks vary, resulting in unique fan vibration patterns. We leverage these specific traits to distinguish and classify various applications (Fig. 5).

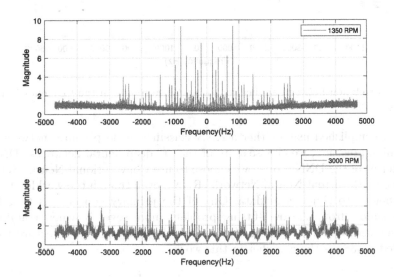

Fig. 5. Frequency components of mmFingerprint responses to different fan speeds.

3.5 Features Extraction and Applications Classification

Numerous methods exist for extracting the features from time-series data. One approach involves the manual extraction of these features by performing signal analysis, such as Fast Fourier Transform (FFT). Another method is to utilize deep neural networks (DNN) for feature extraction. By employing trained DNN layers, we can extract complex features. Different applications are subsequently categorized based on the features extracted from the mmWave vibration traces. To eliminate the necessity for manual feature crafting, we opt for a machine-learning approach to extract features and classify the workload traces. This job

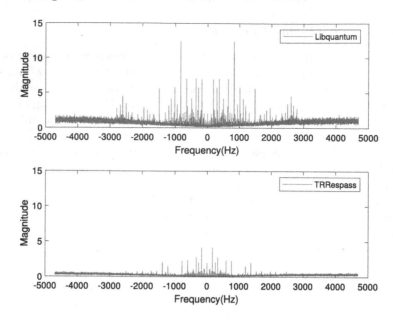

Fig. 6. Correlation between applications from fan response in the frequency domain.

can be accomplished using either a KNN classifier or deep neural networks. To attain high precision, we choose state-of-the-art deep neural networks (DNN).

In terms of the DNN model, we choose to use Convolutional Neural Network (CNN) over Recurrent Neural Network (RNN), even though the workload power traces belong to time series data. One of the primary concerns is that RNN usually suffers from the over-fitting problem more severely when training on long time series [37]. To be specific, we use the ResNet10 architecture that is described in [38] as the classifier in this work.

4   Evaluations

4.1   Experiment Setup

We use a Texas Instruments IWR1642BOOST evaluation board to transmit and receive chirps. The IWR1642 chip can generate chirps with continuous frequency bands of 76 ~ 77 GHz and 77 ~ 81 GHz. The evaluation board integrates two Tx antennas and four Rx antennas. We use the two Tx channels sending out the same FMCW chirps with a continuous band of 3.98 GHz. We use DCA1000EVM evaluation board to extract data samples at a rate of 2.1 Msps. The frame duration is 15 ms with 128 chirps in each frame. The antenna is placed 0.6 m away from the CPU fan with no obstacles in between. The chirps are reflected off the cooling fan and captured by the four Rx antennas. In each case, the positions of the antenna and the target machine are kept constant to eliminate the effects of position movement on the reflected signals.

The targeted machine is equipped with a Gigabyte GA-H170-D3HP motherboard, an i7-6700 CPU, an Intel E97379-003 CPU cooling fan, and an Apevia ATX-SN1050W power supply. The deep learning classifier is built with Keras, using Tensorflow as its backend. This classifier is implemented on a desktop computer powered by an Intel i7-9700K CPU, with 64 GB of DRAM, and an Nvidia RTX3090 GPU.

4.2 Threat Model for Detecting Rowhammer Attempts Using Reputable Tools

Assume an attacker plans to initiate a Rowhammer attempt on a targeted computer system equipped with DDR3 or DDR4, a CPU cooling fan, or a power supply cooling fan. Before the Rowhammer attack, the attacker must scan the memory addresses to determine if the computing systems are vulnerable to Rowhammer attacks. Due to their effectiveness, it is highly likely that the attacker used the most advanced Rowhammer attack tools such as TRRespass and Blacksmith for this reconnaissance process to circumvent the TRR implemented by major vendors in DDR4 DRAMs. Considering the extremely low likelihood of discovering exploitable bit flips within a short time, the attacker would need to scan the DRAM intensively to identify vulnerable bits, recording this information for future exploitation. This step typically requires a significant amount of time. We can set up a millimeter-wave (mmWave) radar at a predetermined distance from the cooling fan of either the CPU or power supply, ensuring that there are no obstructions in the path. This arrangement is feasible for most desktops and servers since their cooling fans are typically visible through ventilation openings. With its high-precision detection capabilities, our system can discern even the smallest variations in the vibrations of the cooling fan during computational processes.

4.3 CPU Cooling Fan Side-Channel

We assess the CPU cooling fan side-channel across various applications, as outlined in Table 1. We select several benign applications and two of the most effective Rowhammer tools against TRR named TRRespass [26] and blacksmith [27]. These benign applications include system idle, playing a video with vlc player, and opening the YouTube webpage. We also evaluate some SPEC 2006 benchmarks including data compression application bzip2 (integer), quantum computation simulator libquantum (integer), playing the game of Go gobmk (integer), fluid dynamics simulation lbm (floating-point), quantum chromodynamics simulationmilc (floating-point). For each workload, we construct 500 individual traces, each lasting 0.96 s with 8192 equivalent samples.

Dataset. The dataset is composed of ten distinct classes, which are divided into training and test sets at a proportion of 80% and 20%, respectively. The deep learning classifier undergoes training for 500 epochs using the training dataset and its performance is subsequently evaluated on the test dataset.

Table 1. Evaluated applications

Label	applications	Notes
0	blacksmith	Rowhammer
1	bzip2	CINT
2	gobmk	CINT
3	idle	
4	lbm	CFP
5	libquantum	CINT
6	milc	CFP
7	TRRespass	Rowhammer
8	vlc	video
9	Youtube	

Evaluation Metrics and Results. The effectiveness of mmFingerprint is assessed using precision, recall, and F1-score as performance measures. The evaluation confusion matrix is presented in Fig. 7(a) and the precision, recall, and F1-score are shown in Table 2. The mmFingerprint has demonstrated an impressive ability to categorize ten distinct classes with an overall accuracy rate of 0.89. Additionally, it exhibits an almost flawless accuracy rate nearing 1.00 when distinguishing two specific Rowhammer tools, data compression bzip2, and playing youtube from other applications. The classifier can recognize gobmk with perfect precision, but a slightly lower recall of 0.93, which has lowered the F1-score to 0.97. This suggests that the model occasionally misses true positives for this class. mmFingerprint has relatively lower precision recognizing idle, lbm, and libquantum, but the model has good recall for these classes. This indicates the model occasionally misclassifies other instances as these classes, but does well in identifying true instances of these classes. The lowest F1-scores on distinguishing milc and vlc, suggesting that the model struggles the most with these classes. When dealing with milc, the model struggles to correctly identify all true instances (recall of 0.59), and for vlc, it frequently misclassifies other instances as this class (precision of 0.92), leading to lower F1-scores. Overall, mmFingerprint performs well on most classes, especially for Rowhammer tools.

4.4 Power Supply Cooling Fan Side-Channel

To assess the efficiency of mmFingerprint when dealing with power supply coiling fan side-channel, we conduct evaluations using the same applications shown in table 1. We collect 500 traces for each workload and they are split into training and test sets at a proportion of 80% and 20%,

mmFingerprint performs well on the power supply cooling fan. The precision, recall, and F1-score are presented in Table 3 and the confusion matrix is shown in Fig. 7(b). A precision of 1.00 means there were no false positive instances. It

Table 2. Evaluation of CPU Fan

Label	precision	recall	f1-score
0	0.99	1	1
1	1	1	1
2	1	0.93	0.97
3	0.69	0.95	0.8
4	0.75	0.85	0.8
5	0.78	0.92	0.84
6	0.79	0.59	0.68
7	1	1	1
8	0.92	0.56	0.69
9	1	1	1

Table 3. Evaluation of Power Fan

Label	precision	recall	f1-score
0	1	1	1
1	1	1	1
2	0.99	1	1
3	1	1	1
4	1	0.99	0.99
5	0.99	1	0.99
6	1	1	1
7	1	1	1
8	1	1	1
9	1	0.99	1

presents an almost absolute accuracy rate nearing 1.00 when classifying blacksmith, data compression bzip2, system idle, milc, TRRespass, and vlc from other applications. It exhibits a slightly lower precision of 0.99 when classifying gobmk and libquantum, which still indicates a high accuracy. Recall measures the ratio of correctly predicted positive instances to all instances that are actually positive. Like precision, a recall of 1.00 indicates a perfect score. All classes have a recall of 1.00, except for lbm and Youtube which have a slightly lower recall of 0.99. Overall, mmFingerprint can recognize different applications with high performance.

5 Related Work

mmWave Sensing. The ability of mmWave sensing to accurately detect microvibrations underscores its effectiveness. It employs high-frequency radar waves, which are adept at identifying minute alterations in the phase or amplitude of reflected signals, enabling the detection of minute displacements, typically associated with vibrations. We summarize the most recent and important findings related to security and privacy, emphasizing the capabilities of mmWave sensing technology.

These applications include speech recovery such as WaveEar [39], through wall sound reconstruction such as Wavesdropper [40], eavesdropping speech of phone call such as mmEve [33], mmSpy [34] mmEcho [32], construction of a Covert Channel using the mmWave sensing of the status of cooling fan [41], lunching a spoofing attack to vehicles [42], user verification for IoT devices [43]. However, to the best of our knowledge, no studies have yet utilized mmWave sensing for the detection of malicious workloads.

Rowhammer. Ever since the inaugural Rowhammer attack [44], the spectrum of these attacks has broadened with numerous variants coming to light. In response, the research community and major DRAM vendors have put forward a wide array of proposed defenses against these diverse Rowhammer onslaughts.

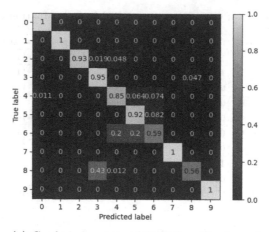

(a) Confusion matrix of CPU fan side-channel.

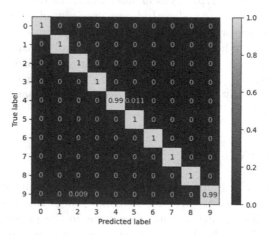

(b) Confusion matrix of power supply fan side-channel.

Fig. 7. Confusion matrix.

The first category is performance counters based Rowhammer detection such as [23, 45, 46]. A second category physically isolates all rows by making only every second row accessible to programs [47]. This method can be circumvented by half-double hammering [48]. Another important way is the Target Row Refresh (TRR) adopted by major DRAM vendors for off-the-share DDR4 DRAMs. This technique is proved to be ineffective for many-sided Rowhammer attacks [26] and half-double hammering [48]. Researchers leveraged EM side-channel to detect the Rowhammer attacks in [12]. But it is unclear whether this can detect the new, sophisticated many-sided hammering and half-double hammering or not.

6 Discussion

In this section, we discuss some situations mmFingerprint can be applied and some limitations.

Although it can not always achieve a 100% detection accuracy, it can significantly improve the detection performance through this new detection method. Moreover, it can complement other existing defense solutions. This system also has the capability to monitor several cooling fans simultaneously. To illustrate, after the application of range FFT, multiple range bins are generated, each corresponding to a specific distance. We can derive varied phase data from these different range bins, which allows us to monitor objects at different distances, thereby observing various cooling fans concurrently. Furthermore, the method presented can potentially be expanded to encompass High-Performance Computing Centers, allowing for the monitoring of illicit applications. An illustration of this would be its application in the detection of unauthorized Cryptocurrency mining activities.

However, certain limitations exist. Detecting minute changes can be challenging, particularly when the execution time is short because the equivalent sampling rate is about 10 kHz with the device we use. Based on the Nyquist sampling theorem, the highest frequency it can sample is less than 5 kHz. The sampling rate is insufficient for capturing applications that have a short execution time, such as those lasting only a few hundred microseconds or less.

7 Conclusion

In our study, we propose a novel application fingerprinting system capable of detecting harmful applications based on the physical side-channel of a cooling fan, specifically focusing on detecting Rowhammer attacks using reputable tools. This system differentiates between the specific characteristics of various applications by utilizing millimeter-wave sensing technology and a machine learning model. Our approach has undergone rigorous assessments, which include evaluations of applications encompassing advanced Rowhammer attack tools like TRRespass and Blacksmith, as well as SPEC2006 benchmarks. These evaluations confirm the high precision of our technique across various scenarios.

Acknowledgment. This work is supported in part by the National Science Foundation (CNS-2147217). The authors would like to thank the anonymous reviewers for their comments and suggestions that help us improve the quality of the paper.

References

1. Wagner, N.R.: Fingerprinting. In: 1983 IEEE Symposium on Security and Privacy, pp. 18–18. IEEE (1983)
2. Yang, L., Zhi, Y., Wei, T., Shui, Yu., Ma, J.: Inference attack in android activity based on program fingerprint. J. Netw. Comput. Appl. **127**, 92–106 (2019)
3. Matyunin, N., Wang, Y., Arul, T., Kullmann, K., Szefer, J., Katzenbeisser, S.: Magneticspy: exploiting magnetometer in mobile devices for website and application fingerprinting. In: Proceedings of the 18th ACM Workshop on Privacy in the Electronic Society, pp. 135–149 (2019)
4. Draghicescu, D., Caranica, A., Vulpe, A., Fratu, O.: Crypto-mining application fingerprinting method. In: 2018 International Conference on Communications (COMM), pp. 543–546. IEEE (2018)
5. Zou, P., Li, A., Barker, K., Ge, R.: Detecting anomalous computation with rnns on gpu-accelerated hpc machines. In: Proceedings of the 49th International Conference on Parallel Processing, pp. 1–11 (2020)
6. Ahmed, M.E., Ullah, S., Kim, H.: Statistical application fingerprinting for ddos attack mitigation. IEEE Trans. Inform. Foren. Sec. **14**(6), 1471–1484 (2018)
7. Singh, S., Estan, C., Varghese, G., Savage, S.: Automated worm fingerprinting. In: OSDI, vol. 4, pp. 4–4 (2004)
8. Chen, Y., Jin, X., Sun, J., Zhang, R., Zhang, Y.: Powerful: mobile app fingerprinting via power analysis. In: IEEE INFOCOM 2017-IEEE Conference on Computer Communications, pp. 1–9. IEEE (2017)
9. Khan, H.A., Sehatbakhsh, N., Nguyen, L.N., Prvulovic, M., Zajić, A.: Malware detection in embedded systems using neural network model for electromagnetic side-channel signals. J. Hardware Syst. Sec. **3**, 305–318 (2019)
10. Liang, S., Zhan, Z., Yao, F., Cheng, L., Zhang, Z.: Clairvoyance: exploiting far-field em emanations of gpu to" see" your dnn models through obstacles at a distance. In: 2022 IEEE Security and Privacy Workshops (SPW), pp. 312–322. IEEE (2022)
11. Zhan, Z., Zhang, Z., Liang, S., Yao, F., Koutsoukos, X.: Graphics peeping unit: Exploiting em side-channel information of gpus to eavesdrop on your neighbors. In: 2022 IEEE Symposium on Security and Privacy (SP), pp. 1440–1457. IEEE (2022)
12. Zhang, Z., Zhan, Z., Balasubramanian, D., Li, B., Volgyesi, P., Koutsoukos, X.: Leveraging em side-channel information to detect rowhammer attacks. In: 2020 IEEE Symposium on Security and Privacy (SP), pp. 729–746. IEEE (2020)
13. Li, J., et al.: {FOAP}:{Fine-Grained}{Open-World} android app fingerprinting. In: 31st USENIX Security Symposium (USENIX Security 22), pp. 1579–1596 (2022)
14. Aceto, G., Ciuonzo, D., Montieri, A., Pescape, A.: Traffic classification of mobile apps through multi-classification. In: GLOBECOM 2017-2017 IEEE Global Communications Conference, pp. 1–6. IEEE (2017)
15. Kolbitsch, C., Comparetti, P.M., Kruegel, C., Kirda, E., Zhou, X., Wang, X.: Effective and efficient malware detection at the end host. In: USENIX security symposium, vol. 4, pp. 351–366 (2009)
16. Seaborn, M., Dullien, T.: Exploiting the dram rowhammer bug to gain kernel privileges. Black Hat **15**, 71 (2015)
17. Xiao, Y., Zhang, X., Zhang, Y., Teodorescu, R.: One bit flips, one cloud flops: cross-vm row hammer attacks and privilege escalation. In: 25th {USENIX} Security Symposium {USENIX} Security 2016, pp. 19–35 (2016)

18. Bhattacharya, S., Mukhopadhyay, D.: Curious case of Rowhammer: flipping secret exponent bits using timing analysis. In: Gierlichs, B., Poschmann, A.Y. (eds.) CHES 2016. LNCS, vol. 9813, pp. 602–624. Springer, Heidelberg (2016). https://doi.org/10.1007/978-3-662-53140-2_29

19. Jang, Y., Lee, J., Lee, S., Kim, T.: Sgx-bomb: locking down the processor via rowhammer attack. In: Proceedings of the 2nd Workshop on System Software for Trusted Execution, pp. 1–6 (2017)

20. Mutlu, O.: The rowhammer problem and other issues we may face as memory becomes denser. In: Design, Automation & Test in Europe Conference & Exhibition (DATE), vol. 2017, pp. 1116–1121. IEEE (2017)

21. Gruss, D.: Another flip in the wall of rowhammer defenses. In: 2018 IEEE Symposium on Security and Privacy (SP), pp. 245–261. IEEE (2018)

22. Kwong, A., Genkin, D., Gruss, D., Yarom, Y.: Rambleed: reading bits in memory without accessing them. In: 2020 IEEE Symposium on Security and Privacy (SP), pp. 695–711. IEEE (2020)

23. Aweke, Z.B.: Anvil: software-based protection against next-generation rowhammer attacks. ACM SIGPLAN Not. **51**(4), 743–755 (2016)

24. Lee, E., Kang, I., Lee, S., Suh, G.E., Ahn, J.H.: Twice: preventing row-hammering by exploiting time window counters. In: Proceedings of the 46th International Symposium on Computer Architecture, pp. 385–396 (2019)

25. Park, Y., Kwon, W., Lee, E., Ham, T.J., Ahn, J.H., Lee, J.W.: Graphene: strong yet lightweight row hammer protection. In: 2020 53rd Annual IEEE/ACM International Symposium on Microarchitecture (MICRO), pp. 1–13. IEEE (2020)

26. Frigo, P.: Trrespass: exploiting the many sides of target row refresh. In: 2020 IEEE Symposium on Security and Privacy (SP), pp. 747–762. IEEE (2020)

27. Jattke, P., Van Der Veen, V., Frigo, P., Gunter, S., Razavi, K.: Blacksmith: scalable rowhammering in the frequency domain. In: 2022 IEEE Symposium on Security and Privacy (SP), pp. 716–734. IEEE (2022)

28. UEFI Forum, Inc., Advanced configuration and power interface (acpi) specification (2022). https://uefi.org/sites/default/files/resources/ACPI_Spec_6_5_Aug29.pdf. (Accessed 06 Dec 2023)

29. Hanrahan, D.: Fan-speed control techniques in pcs. Analog Dialogue **34**(4), 34–04 (2000)

30. Bilik, I., Longman, O., Villeval, S., Tabrikian, J.: The rise of radar for autonomous vehicles: signal processing solutions and future research directions. IEEE Signal Process. Mag. **36**(5), 20–31 (2019)

31. Li, X., Wang, X., Yang, Q., Song, F.: Signal processing for tdm mimo fmcw millimeter-wave radar sensors. IEEE Access **9**, 167959–167971 (2021)

32. Hu, P., Li, W., Spolaor, R., Cheng, X.: mmecho: a mmwave-based acoustic eavesdropping method. In: 2023 IEEE Symposium on Security and Privacy (SP), pp. 836–852. IEEE Computer Society (2022)

33. Wang, C.: mmeve: eavesdropping on smartphone's earpiece via cots mmwave device. In: Proceedings of the 28th Annual International Conference on Mobile Computing and Networking, pp. 338–351 (2022)

34. Basak, S., Gowda, M.: mmspy: spying phone calls using mmwave radars. In: 2022 IEEE Symposium on Security and Privacy (SP), pp. 1211–1228. IEEE (2022)

35. Jiang, C., Guo, J., He, Y., Jin, M., Li, S., Liu, Y.: mmvib: micrometer-level vibration measurement with mmwave radar. In: Proceedings of the 26th Annual International Conference on Mobile Computing and Networking, pp. 1–13 (2020)

36. Rao, S.: Introduction to mmwave sensing: Fmcw radars. Texas Instruments (TI) mmWave Training Series, pp. 1–11 (2017)

37. Fawaz, H.I., Forestier, G., Weber, J., Idoumghar, L., Muller, P.-A.: Deep learning for time series classification: a review. Data Mining Knowl. Dis. **33**(4), 917–963 (2019)
38. Wang, Z., Yan, W., Oates, T.: Time series classification from scratch with deep neural networks: a strong baseline. In: 2017 International joint conference on neural networks (IJCNN), pp. 1578–1585. IEEE (2017)
39. Xu, C.: Waveear: exploring a mmwave-based noise-resistant speech sensing for voice-user interface. In: Proceedings of the 17th Annual International Conference on Mobile Systems, Applications, and Services, pp. 14–26 (2019)
40. Wang, C., Lin, F., Ba, Z., Zhang, F., Wenyao, X., Ren, K.: Wavesdropper: through-wall word detection of human speech via commercial mmwave devices. Proc. ACM Interac. Mobile, Wearable Ubiquitous Technol. **6**(2), 1–26 (2022)
41. Li, Z.: Spiralspy: exploring a stealthy and practical covert channel to attack air-gapped computing devices via mmwave sensing. In: The 29th Network and Distributed System Security (NDSS) Symposium 2022. The Internet Society (2022)
42. Vennam, R.R.: mmspoof: resilient spoofing of automotive millimeter-wave radars using reflect array. In: 2023 IEEE Symposium on Security and Privacy (SP), pp. 1971–1985. IEEE Computer Society (2022)
43. Dong, Y., Yao, Y.-D.: Secure mmwave-radar-based speaker verification for iot smart home. IEEE Internet Things J. **8**(5), 3500–3511 (2020)
44. Kim, Y.: Flipping bits in memory without accessing them: an experimental study of dram disturbance errors. ACM SIGARCH Comput. Architecture News **42**(3), 361–372 (2014)
45. Gruss, D., Maurice, C., Wagner, K., Mangard, S.: Flush+Flush: a fast and stealthy cache attack. In: Caballero, J., Zurutuza, U., Rodríguez, R.J. (eds.) DIMVA 2016. LNCS, vol. 9721, pp. 279–299. Springer, Cham (2016). https://doi.org/10.1007/978-3-319-40667-1_14
46. Zhang, T., Zhang, Y., Lee, R.B.: CloudRadar: a real-time side-channel attack detection system in clouds. In: Monrose, F., Dacier, M., Blanc, G., Garcia-Alfaro, J. (eds.) RAID 2016. LNCS, vol. 9854, pp. 118–140. Springer, Cham (2016). https://doi.org/10.1007/978-3-319-45719-2_6
47. Konoth, R.K.: Zebram: comprehensive and compatible software protection against rowhammer attacks. In: 13th {USENIX} Symposium on Operating Systems Design and Implementation {OSDI} 2018, pp. 697–710 (2018)
48. Kogler, A., et al.: {Half-Double}: Hammering from the next row over. In: 31st USENIX Security Symposium (USENIX Security 2022), pp. 3807–3824 (2022)

ADC-Bank: Detecting Acoustic Out-of-Band Signal Injection on Inertial Sensors

Jianyi Zhang[1,3]([📧]), Yuchen Wang[2], Yazhou Tu[3], Sara Rampazzi[4], Zhiqiang Lin[5], Insup Lee[6], and Xiali Hei[3]

[1] Beijing Electronic Science and Technology Institute, Beijing 100070, BJ, China
zjy@besti.edu.cn

[2] Academy of Information and Communications Technology, Beijing 100191, BJ, China

[3] University of Louisiana at Lafayette, Lafayette, LA 70503, USA
{yazhou.tu1,xiali.hei}@louisiana.edu

[4] University of Florida, Gainesville, FL 32611, USA
srampazzi@ufl.edu

[5] Ohio State University, Columbus, OH 43210, USA
zlin@cse.ohio-state.edu

[6] University of Pennsylvania, Philadelphia, PA 19104, USA
lee@cis.upenn.edu

Abstract. Inertial sensors are widely used in navigation, motion tracking, and gesture recognition systems. However, these sensors are vulnerable to spoofing attacks, where an attacker injects a carefully designed acoustic signal to trick the sensor readings. Traditional approaches to detecting and mitigating attacks rely on module redundancy, i.e., adding multiple sensor modules to increase robustness. However, this approach is not always feasible due to the limited space and increased complexity of current printed circuit boards.

This paper proposes a new method, ADC-Bank, to detect inertial sensor spoofing attacks via acoustic out-of-band signals. Unlike other multiple-sensor-based solutions, it is based on component redundancy within one sensor, using multiple analog-to-digital converters (ADCs) with different sampling rates to simultaneously sample the output of the sensors. The different sample rates result in different aliasing frequencies for out-of-band signals that can be used to detect attacks. The proposed method is evaluated on off-the-shelf inertial sensors with commercial ADCs, demonstrating its ability to detect the attacking signals with relatively low cost and computation overhead.

Keywords: Spoof attack · Out-of-band acoustic signal injection · Inertial sensor · Multiple ADCs · Detection and correction

1 Introduction

Micro-electro-mechanical systems (MEMS) inertial sensors are known to be susceptible to acoustic out-of-band signal injections [4–6,9,30,36–38,40,41]. These

© ICST Institute for Computer Sciences, Social Informatics and Telecommunications Engineering 2024
Published by Springer Nature Switzerland AG 2024. All Rights Reserved
Y. Chen et al. (Eds.): SmartSP 2023, LNICST 552, pp. 53–72, 2024.
https://doi.org/10.1007/978-3-031-51630-6_4

attacks used acoustic signals at frequencies close to the sensor's resonant frequency to induce high-frequency analog signals in the sensor circuits. Ideally, the injected signals should be filtered out because they are out-of-band signals. However, attackers can still inject these signals into the system. The essential feature of out-of-band signal injections is that the induced analog signals will be undersampled, resulting in signal aliasing. When aliasing occurs, attackers can change the output of sensors by maliciously generated stimuli, then deceive the sensing and actuation systems into executing malicious actions accordingly [11]. For example, a self-balancing scooter can adjust direction and speed according to its lean angles, which are described by inertial sensors. However, an attacker can induce an intentional sound at the resonant frequencies of the gyroscope; the output of the inertial sensor will be distorted, and the attacker can make the scooter move in a corresponding opposite direction [38].

In recent years, several defense strategies have been studied to solve the problem of acoustic-based spoofing attacks. For example, shielding [3,16,30,41] was recommended to mitigate out-of-band injections into inertial sensors. However, shielding can cause heat dissipation, cost, size, and usability issues. Another defense consists of low-pass filters that can filter out malicious high-frequency signals and mitigate attack at inertial sensors [16,37,45]. In practice, implementing ideal anti-aliasing filters that eliminate all out-of-band signals is trivial. For example, a high-order filter that eliminates all signals above the cutoff frequency will cause signals that change rapidly to ring on for a long time. Moreover, analog filters lead to an unequal time delay as a function of frequency [33]. If the phase delay introduced by filters is large, it is difficult to minimize this delay or compensate for it in software [8]. Moreover, the integrated low-pass filter does not have clear cut-offs [25,32]. An additional defense approach consists of using high-frequency sampling of the analog signal. For instance, the inertial sensor signal frequency induced by movement is generally below 20 Hz. If the sensor designers choose ADCs with sampling rates high enough to handle the resonant frequencies, it will increase the production costs and decrease the sampling resolution and the processing speed due to the over-wide bandwidth. Recent work has studied purely software-based detection methods [35] and module redundancy methods (multisensors for sensor fusion) [3,19,28,41–43]. However, false positives/negatives can occur when external factors or injected data differ from the assumed patterns. The researchers also noted that attacks with a directed magnetic field that can precisely control both the magnetometer and the gyroscope would cause their sensor fusion-based detection method to fail [35].

In this paper, we present ADC-Bank, a novel out-of-band signal defense method using component redundancy within a sensor in contrast to the work mentioned above. Compared to other defense strategies, our method is easy to manufacture and has fewer attack surfaces than module redundancy strategies, such as multiple sensor-based methods. After implementing multiple circuit components that simultaneously elaborate the physical stimulus under different configurations and settings, we provide multiple metrics on the legitimacy of the measurement at the software layer. This information is then used to detect the system from processing an altered signal. We evaluate our method on off-the-shelf inertial MEMS sensors from three different vendors. Our experimental results show that ADC-Bank can detect physical injection attacks via out-of-band acoustic signals on all models of inertial MEMS sensors that we tested.

Despite many existing defense mechanisms against acoustic physical injection attacks at MEMS sensors, there is no fundamental solution to detect these malicious transmissions and prevent vulnerabilities in the physics of a MEMS sensor. Our work fills this gap through the following contributions:

1. We propose a component redundancy scheme to detect acoustic out-of-band signal injection by elaborating and comparing the physical stimulus in different settings.
2. We investigate how to extract the real physical stimulus from different results of the redundant components.
3. We deploy our defense method on off-the-shelf inertial sensors with commercial ADCs to evaluate our method.
4. We discuss how our strategy can be used in the design and manufacturing of future sensors.

2   Background

2.1   MEMS Inertial Sensors

Almost all MEMS inertial sensors have a mass and a support spring, and they use this mechanical structure to detect motion stimuli [26]. MEMS accelerometers sense linear accelerations by displacement of the mass supported by springs and measure the capacitance change between the mass and fixed electrodes [17,44]. MEMS gyroscopes are relatively complex. They have a continuously vibrating mass that, like accelerometers, is supported by springs. They measure the Coriolis force generated by the applied angular velocity on the vibrating mass [29].

After transduction, the sensor output needs a series of additional processing to interface with external components such as microcontrollers. In general, the change in capacitance causes a change in voltage. For an analog sensor, this analog signal is typically amplified and outputted directly from the amplifier. For the digital sensor, the amplified signal is digitized via an analog-to-digital converter (ADC) and then transferred to the control system by standard digital interfaces like SPI, I^{2}C, and UART. In this work, we consider analog inertial sensors to explain our approach.

2.2   ADCs and Aliasing

After the sensor transforms the physical measurement into an analog signal, a built-in ADC digitizes the sensor's output. The analog signal that is continuous in time should be converted at a certain rate by ADC, and this rate is defined as the sampling rate or sampling frequency of the converter. According to the Nyquist-Shannon sampling theorem, the sampling rate should be at least twice the signal's maximum frequency when the original physical measurement can be reconstructed from the discrete data by the mitigation filter. If the system acquires data at an insufficient rate, called undersampling, the signal will be incorrectly detected at a specified interval as a lower frequency [18]. Then aliasing will occur.

2.3 Acoustic Injection

Because of its miniaturized mechanical and integrated electronic structure, these sensors' output could be changed to incorrect values by resonant acoustic interferences [37]. The successful modification relies on two vulnerabilities of the MEMS inertial sensor: the mass-spring structure that works as the receiving system for resonant acoustic signals and the non-linearity of electronic components like the overdriven amplifier or under-sampling of an ADC. According to the second vulnerability, the acoustic injection attack can be categorized into two classes: output control attack and output biasing attack [37]. The output control attack leverage signal clipping at the insecure amplifier to introduce a DC component into the acceleration signal, which slips through any subsequent LPF [15,27,39]. However, triggering this kind of attack requires a signal beyond the amplifier's capability, which means high power and deafening volume. Therefore, it becomes impractical to generate the required loudness and attack the sensor from a long distance [10].

3 Threat Model

We assume that the attackers' objective is to spoof and manipulate the MEMS inertial sensors' output. To achieve this, attackers need to transmit specific acoustic signals at the resonant frequencies to deceive the sensor and trigger the control system's actuation.

Attack Scenarios. We assume that attackers can use an off-the-shelf speaker or transducer to generate the sound waveforms for the injection. Also, we assume that they are able to induce the sound, at the resonant frequencies of these sensors, at any position, distance, or angle. This might be done via means of amplifiers and constant directivity horns. We assume that the attackers have sufficient resources to optimize the power, directivity, and emitting area. More powerful attackers may utilize customized acoustic equipment to improve the effect. The signal source of attacks can be a built-in speaker, a function generator, an MCU board like Arduino, mini signal generator boards [24,31], or even malicious codes in an email or webpage with JavaScript and autoplay audio enabled. The attacker can also use long-distance acoustic devices to play the sound waves as described by Tu et al. [38].

Attack Goals. We assume that attackers utilize the resonant acoustic signal to inject the sensor output and deliver adversarial control to the system. Such

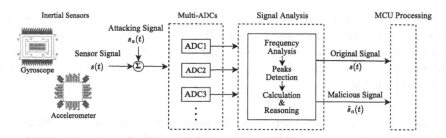

Fig. 1. Scheme of an ADC-Bank's signal processing. The measured output of the sensor is a linear combination of the original signal, $s(t)$, and the injection signal, $s_a(t)$. Unlike legitimate signals, malicious out-of-band signals sampled by the ADCs generate multiple frequency peaks. We can employ such an observation to detect and analyze various attack scenarios.

attacks on the IMU sensors will pose security and safety risks to cyber-physical systems like robots, stabilization systems, self-balancing scooters, drones, etc.

System Accessibility. We assume the attackers know the exact type and model of the MEMS inertial sensors and can easily access the datasheet to know the sensors' components and structures.

4   Defense Approach

4.1   System Model

The system model of our proposed protection scheme is presented in Fig. 1. It has two blocks, including a multi-ADCs part and a signal analysis module.

The multi-ADCs part consists of more than two ADCs whose sampling rates have certain constraints like pairwise relatively prime. After sampling the sensor output synchronously, these ADCs send their respective measurement results of the same sensor to the signal analysis module.

The signal analysis module for spectral analysis consists of three parts: frequency analysis, peak detection, and a calculation and reasoning phase. The frequency analysis performs a Fast Fourier transform (FFT) on each measurement result of the different ADCs and transfers the detection into the frequency domain. According to the results of peak detection, ideally, there will be one overlapped peak in the frequency domain, which means that the signal has normal behavior. Otherwise, multiple separated peaks would suggest the presence of out-of-band physical signal injection attacks. In the calculation and reasoning phase, when no multiple peaks are found, which means that there is no injection signal, the A/D conversion and the measurement value are considered trustworthy. Hence, the actuation system knows the result is digitized from the original sensor's output. However, when the signal injection attack is detected, we can calculate the approximate frequencies of the injection signal based on prior knowledge of the intended frequency range of the attack (more details in [38]).

From the signal perspective, the sensor output generated by the real movement is $s(t)$ in the absence of attackers. The attack signal is $s_a(t)$ generated by acoustic injection. We model the measured output of the sensor as a linear combination of the original signal $s(t)$ and the injection signal $s_a(t)$. Hence the measured signal $\widetilde{s}(t)$ is:

$$\widetilde{s}(t) = s(t) + s_a(t) \tag{1}$$

Since the mechanical structure of the sensor under resonance oscillates at the same frequency as the attacking signal, we model the resulting signal with a resonant frequency F_a and an initial phase ϕ as:

$$s_a(t) = A \cdot sin(2\pi F_a t + \phi) \tag{2}$$

where coefficients $A = A_0 k_a k_s$. A_0 is the amplitude of the attacking signal, k_a is the acoustics attenuation when the attacking signal is transmitted to the target sensor, and k_s is the sensitivity of the sensing mass. Substitute Eq. (2) into Eq. (1), we have the measured value:

$$\widetilde{s}(t) = s(t) + A \cdot sin(2\pi F_a t + \phi) \tag{3}$$

Then, the combination signal will be sampled by multiple ADCs. Typically, the sampling rate of the ADC in the inertial sensor system is designed to be high enough to sample the movement signal, so the true sensor measurement $s(t)$ will be normally converted. However, the frequency of attacking signals injected through resonance is usually much higher than the sampling rate. Therefore, sampling these out-of-band high-frequency signals will cause aliasing. A sinusoidal analog signal with frequency F will be aliased to a digital signal with a frequency of ε when $F > 2F_S$, where F_S is the sampling rate. We have

$$F = n \cdot F_S + \varepsilon \quad (-\frac{1}{2}F_S < \varepsilon \leq \frac{1}{2}F_S, \; n \in \mathbb{Z}^+) \tag{4}$$

Therefore, assuming that F_a is the resonant frequency of the sensor, the adversary uses it as the frequency of injection signals. For multiple ADCs, based on Eq. (4), we have:

$$F_a = n_i \cdot F_{Si} + \varepsilon_i \quad (-\frac{1}{2}F_{Si} < \varepsilon_i \leq \frac{1}{2}F_{Si}, n_i \in \mathbb{Z}^+) \tag{5}$$

where F_{Si} is the sampling rate of the i-th ADC, and ε_i is the resulted frequency of the corresponding ADC output. For simplicity, we assume that n in Eq. (4) and Eq. (5) is the integer multiple of the sampling rate F_S. Therefore, n_i stays the same when ε, F_S changes slightly.

According to Eq. (5), these multiple ADCs with different sampling rates will generate different results ε_i for the same input signal F_S. Out-of-band signal injections can be detected on the basis of this separation. Meanwhile, based on these sampling rates, the possible n_i can be traversed according to the reading, and the approximate range of F_a in Eq. (5) can be found according to multiple F_{Si} and ε_i.

5 Detection

Multi-ADCs. In the out-of-band signal injection attack against the MEMS inertial sensor, under acoustic injection, malicious sound waves are transmitted to the mechanical structure of the inertial sensor, forcing the sensing mass to resonate.

In the analog-to-digital conversion process, the input signal is sampled. Only when the sampling rate F_S is greater than twice the highest frequency $2F_{Max}$ in the analog signal spectrum, the analog signal can be recovered without distortion. Therefore, the ADC sampling rate in the inertial sensor system is designed to be high enough to sample the movement signal. However, when inertial sensors face ultrasound resonant signal injection, also known as out-of-band signal injection attacks, the frequency of attacking signals is usually much higher than the sampling rate.

The sampling rate in the inertial sensor system is usually in the tens or hundreds, while the resonant frequency is usually higher than 2 kHz for the accelerometer and 19 kHz for the gyroscope. Since the resonant frequency is much higher than the sampling rate, signal aliasing will occur and be reconstructed into a new low-frequency in-band signal.

To detect suspicious out-of-band signal injection attacks, we take advantage of the phenomenon of undersampling. Specifically, the multi-ADC part consists of more than two ADCs that sample the input signal, respectively. Then, reconstructing these undersampled signals from the digital samples will cause signal aliasing. Our defense solution consists of comparing such aliased signals to determine the reconstructed original signal.

The microcontroller of the control system can then be used to measure the physical quantity and hence can detect the attack based on the outputs of the ADCs. We suppose that the attacker remotely injects the malicious waveforms into the inertial sensor circuit. After sampling and digitizing the stimulus by multiple ADCs with different sampling rates, the control system can spot the attack immediately since the results in the frequency domain are totally different. With the help of well-designed parameters, we can not only detect the existence of malicious signals, but also recover the real signal from the measurement of the sensor's outputs. In particular, if the sampling rates of multiple ADCs we selected are pairwise relatively prime, according to the Chinese remainder theorem [21], the microcontroller can easily calculate the range of attack frequencies. After that, we can easily filter the frequencies induced by attack signals and provide reliable measurements to the control system.

Frequency Analysis. In our defense approach, a key part is to analyze the frequency of the reconstructed signal. When multi-ADCs sample and digitize the input signal respectively, each measurement result of the different ADCs will be performed frequency analysis via Fast Fourier transform (FFT) and transferred the detection into the frequency domain. With the help of frequency domain analysis, we can obtain the frequency information of the input signal immediately.

If the input signal is generated by normal motion, the maximum frequency of the input signal must be within half of the sampling rate of the ADCs, and it will be able to be digitized normally. This means that the measurement results of different ADCs will produce the same frequency component after FFT. When faced with the injection of malicious acoustic signals, the situation will become different. The input signals far beyond the sampling rate of ADCs will cause aliasing. Because the sampling rates of ADCs are different from each other and relatively prime to each other, the measurement results will produce different frequencies after FFT. We use a simple peak detection algorithm to determine the credibility of the measured sensor value.

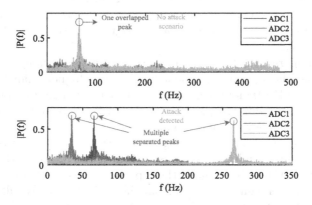

Fig. 2. Scheme of an ADC-Bank's attack detection. In contrast with legitimate signals, malicious out-of-band signals sampled by the ADCs generate multiple frequency peaks. This technique can be used to detect and analyze various attack scenarios.

Peak Detection. The peak detection algorithm is used to quickly measure the results after FFT. Figure 2 shows the main peak detection process in the frequency domain; we detect the peak of the FFT results of ADC measurements, respectively. If only one overlapping peak is detected, it indicates that the signal is not attacked and is credible for the subsequent actuation system. If there are multiple separate peaks, that means that there is a potential attack. These signals will not be able to be transmitted directly to the actuation system and will need to be corrected.

6   Experiments

In this section, we try to prove the effectiveness of our method in a real-world case study. To prove the effectiveness of our signal process scheme, we designed a series of experiments. We have built an acoustic injection attack environment to collect raw data and perform signal processing and analysis.

We evaluate our approach from the following two situations. 1) To simulate a real attack environment, we use a signal generator output to drive the speaker and then interfere with the inertial sensor. Then we collect the motion signal and the injection signal from the inertial sensor, respectively, using an NI USB-4431 Data Acquisition (DAQ) [14]. 2) We use ADCs and microcontroller units (MCUs) to build a set of data acquisition environments. We use a signal generator output to drive the speaker, and then interfere with the inertial sensor. Then we collect and upload the sensor data to a PC for further signal processing.

6.1   Experimental Setup

Figure 3 shows the experimental setup. DG5300 signal generator is used to generate an acoustic signal [23]. Here, the output amplitude is set to 5 v. A power amplifier is used to enhance signal power, and the Vifa speaker [2] is responsible for outputting acoustic waves. The inertial sensor chip is mounted on an evaluation board and driven by 3.3 v/5 v DC provided by the external Arduino [1].

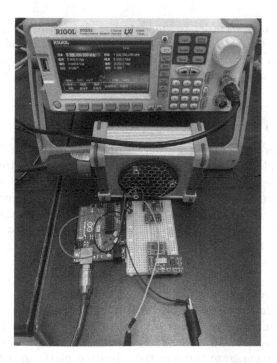

Fig. 3. Schematic of the experimental setup. The inertial sensor chip is mounted on an evaluation board placed on the experimental platform. The attack range can be between one and three meters in a real attack scenario [38].

After determining the resonant frequency of each inertial sensor chip, we select the appropriate frequency to carry out an acoustic injection attack on each chip and then use the NI USB-4431 DAQ module and the A/D Data Acquisition System we built to collect the senors' output, respectively.

In the following experiments, ADXL335 is used as the target accelerometer [7], and LPY550AL is used as the target gyroscope [34]. We also selected some other inertial sensor models, as shown in Table 1.

Table 1. Resonant frequency and aliasing frequency results of inertial sensors in the experiment

Chip Enterprise	Chip Model	Type	Axis	Resonant Frequency	Attack Frequency	Aliasing Frequency				
						NI USB-4431 DAQ Results			A/D Results	
						280 Hz	700 Hz	1000 Hz	250 Hz	920 Hz
Murata	ENC-03MB	Gyro	x	22 kHz 25.2 kHz	25135 Hz	215 Hz	65 Hz	135 Hz	115 Hz	295 Hz
Murata	ENC-03RC	Gyro	x	30 kHz–33 kHz	32295 Hz	185 Hz	95 Hz	295 Hz	45 Hz	95 Hz
STMicroelectronics	LPY550AL	Gyro	x	22 kHz–23 kHz	22785 Hz	105 Hz	315 Hz	215 Hz	35 Hz	215 Hz
STMicroelectronics	LPY550AL	Gyro	y	22 kHz–23 kHz	–	–	–	–	–	–
ADI	ADXL335	Acce	x	4 kHz–5.5 kHz	4490 Hz	10 Hz	290 Hz	490 Hz	10 Hz	110 Hz
ADI	ADXL335	Acce	y	4 kHz–5.5 kHz	–	–	–	–	–	–
ADI	ADXL335	Acce	z	4 kHz–5.5 kHz	–	–	–	–	–	–

At the same time, in order to simulate the signal output generated by real motion, we place the inertial sensor chip mounted on an evaluation board on top of a vibration platform, where we set the vibrating frequency below 50 Hz, then we use the above two acquisition systems to collect the signal output, respectively.

6.2 Evaluation Experiment

Inertial Sensors with DAQ. In this set of experiments, we first put the chip on the vibration platform and then set the vibration platform frequency to 16 Hz to simulate a true motion. For each time sensor output, we sample the output using three different sampling rates and analyze it in the frequency domain.

Firstly, we carried out experiments on an accelerometer, ADXL335. Figure 4 shows the detailed results of ADXL335. Figure 4a shows the raw data sampled by the system in the time domain. We process the data before transforming them to the frequency domain. First, we remove the DC component of the signal because it has no significance for us in detecting the frequency of the sensor output signal. Second, we normalize the data to make their amplitudes close. Then we transform the data sampled at three different sampling rates into the frequency domain, and the results are shown in Fig. 4b. We found that there are overlapping peaks at 16 Hz in the spectrum. This is also consistent with the frequency that we generate through the vibrating platform, which means that this signal is a normal motion signal. Our signal processing scheme will give the signal high confidence and let the signal enter the control system without affecting the response of the actuation system.

Then, we conducted a similar experiment on a gyroscope, LPY550AL. The environment settings are the same as ADXL335. The results are shown in Fig. 5. Similarly, we can also see that in Fig. 5b), there is an overlapping peak at 16 Hz.

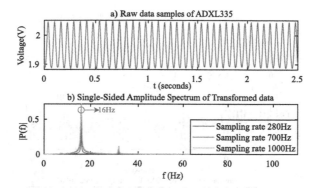

Fig. 4. The testing results of ADXL335 with NI USB-4431 DAQ. a) shows the sampled time domain raw data. In b), the raw time domain data is converted to the frequency domain after data processing.

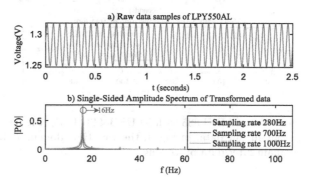

Fig. 5. The testing results of LPY550AL with NI USB-4431 DAQ. a) shows the sampled time domain raw data. In b), the raw time domain data is converted to the frequency domain after data processing.

Since the NI USB-4431 DAQ module has a good anti-aliasing filter, the resonant signal beyond the sampling rate will not be collected. Therefore, we choose a sampling rate of 70,000 Hz, which is much higher than the resonant frequency and can sample normally, and then we simulate the aliasing process under different sampling rates by down-sampling.

For ADXL335, we use the signal generator to generate a 3,525 Hz signal, which is also the resonant frequency of the inertial sensor. The signal is output through the speaker to interfere with the accelerometer. By down-sampling, we get the raw data sampled at three sampling rates. After the signal is processed, we convert it to the frequency domain.

The data acquisition and frequency analysis results are shown in Fig. 6. Figure 6 a), b), and c) are the raw time-domain data of resonant signal down-sampled to 280 Hz, 700 Hz, and 1,000 Hz (due to the limitation of down-sampling), respectively. Figure 6d) is data converted to the frequency domain

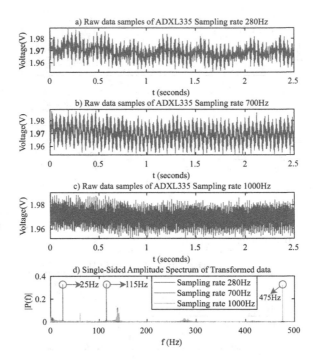

Fig. 6. The testing results of ADXL335 with NI USB-4431 DAQ. Figure a, b and c show the sampled time domain raw data. In Fig. d, the raw time domain data is converted to the frequency domain after data processing.

after signal processing. We can clearly see in the spectrum that there are three separate peaks because different sampling rates lead to different aliasing frequencies. Therefore, we can determine that this abnormal signal needs filtering.

Regarding the gyroscope, we also performed experiments on an LPY550AL. We generated a 22,700 Hz signal, which is also the resonant frequency of the inertial sensor, through the signal generator and output by a speaker. The results are shown in Fig. 7. In the spectrum diagram, we can see that there are only two separated peaks, one of which is the overlapping peak, which is due to the same aliasing frequency of the two sampling rates. However, it can still be determined that the signal is abnormal.

In this set of experiments, we also tested other types of inertial sensor chips, and the results are shown in Table 1. In fact, on all types of inertial sensor chips, we can clearly distinguish whether there is abnormal signal input. This also preliminarily proves that our detection method is applicable to real-world actuation systems.

Inertial Sensors with commercial ADCs. The main difference between this group of experiments and the previous group of experiments lies in the sampling method. We did not use a professional DAQ module to collect the inertial sensor

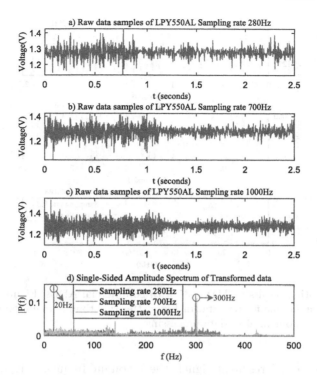

Fig. 7. The testing results of LPY550AL with NI USB- 4431 DAQ. Figure a, b and c show the sampled time domain raw data. In Fig. d, the raw time domain data is converted to the frequency domain after data processing.

output like before; instead, we will use a commercial ADC to collect the inertial sensor output to fully simulate the situation in a real actuation system.

In this set of experiments, we also tested four different inertial sensor models, as shown in Table 1, including the acquisition and analysis of normal motion signals and abnormal resonant signals. The ADC model we use is ADS1015, which has seven optional sampling rates. For normal motion signals, the sampling rate is usually several hundred Hz. At the same time, to evaluate whether the two ADCs can completely detect abnormal signals, we use two ADS1015, which are connected to the Arduino microcontroller and configured with different sampling rates. Here, we set the sampling rates of ADC to 250 Hz and 920 Hz (which are two optional sampling rates for ADS1015), respectively. We use these two ADCs with different sampling rates to sample the inertial sensor output simultaneously, then upload the sampling data to a PC for further data processing and analysis.

Here, we take ADXL335 as an example, as shown in Fig. 8. For the normal motion signal, we also select 16 Hz as the vibration frequency to simulate normal motion. We observed overlapping peaks at 16 Hz in Fig. 8c. It can be seen that the in-band, normal motion signals can be determined to be trustworthy by two ADCs' simultaneous sampling.

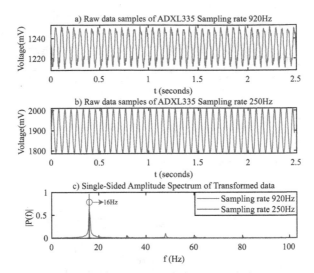

Fig. 8. The testing results of ADXL335 with commercial ADCs. Figure a and b show the sampled time domain raw data. In Fig. c, the raw time domain data is converted to the frequency domain after data processing.

For an abnormal resonant signal, the resonant frequency is 4,485 Hz. The results of signal sampling are shown in Fig. 9. As shown in Fig. 9c, two different peaks are generated at two different sampling rates. In fact, there is a certain deviation between the peak frequency and the theoretical aliasing frequency, but we can still determine that there is an abnormal signal to be filtered.

6.3 Attack Frequency Analysis

During the previous data acquisition and processing, we set the attack signal frequency through the signal generator, and then obtain the ADC sampling rate and the frequency of an aliased in-band signal. According to the prior knowledge, we have a known range of possible attack frequencies. For the sampling rate of each ADC, we can traverse the possible small frequency ranges within the possible attack frequency range according to the in-band signal frequency. Then we find the intersection of the frequency ranges determined by different ADCs, and can obtain a calculated attack frequency range. Through the previous experiments, we have collected data from some models of inertial sensors. Next, we will calculate and analyze the specific data for example.

According to Eq. (5), the possible attack frequency ranges of several segments can be calculated according to the peak frequency obtained from ADC of a certain sampling rate. We have a prior range of attack frequency, 2–5 kHz for accelerometers and 19–27 kHz for gyroscopes. For multiple ADCs, we can find different ranges of their peak frequencies, and get the intersections of these ranges. Under the above experimental configuration, we try to calculate the

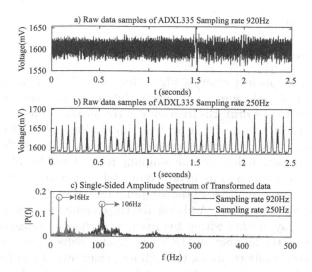

Fig. 9. The testing results of ADXL335 with commercial ADCs. Figure a and b show the sampled time domain raw data. In Fig. c, the raw time domain data is converted to the frequency domain after data processing.

attack frequency range according to the aliasing frequency and ADC sampling rate. We show the results as shown in Fig. 10.

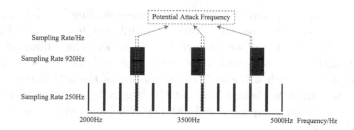

Fig. 10. Frequency range determined by calculation. The black part in the figure is the possible attack frequency range calculated within the range of 2–5 kHz. The red line indicates the overlapping range calculated under the sampling rates of 250 Hz and 920 Hz. We regard the overlapping range as the potential attack frequency range. (Color figure online)

As shown in Fig. 10, we can see that the attack frequency ranges determined by multiple ADCs form an intersection. We compare the attack frequency set by the signal generator with the calculated results, and it can be seen that the actual attack frequency falls within the frequency range we calculated.

7   Discussion

7.1   Adaptive Attacks and Frequency Drift

In acoustic-based spoofing attacks, slight frequency drift or sample rate jitter could be amplified and cause significant deviation in the digital output of the sensors [38]. Due to this drift, the frequency of the aliased output is not constant.

During adaptive attacks, if the attacker knows the details of the algorithm and the sampling rate, and hopes to attack with acoustic signals whose frequency is the common integer multiple of the sampling rates, it will be difficult to implement because the sampling rate is not completely accurate. Even if the attack frequency is an integer multiple of the sampling rate, it will be recognized due to frequency drift in a short time. Additionally, the common integer multiple of the sampling rates may not fall within the resonant frequency range of the inertial sensor. Increasing the number of ADCs will greatly reduce this possibility.

If the attacker wants to attack through frequency sweep and frequency hopping, the attack cannot be implemented because the accurate sampling rate cannot be known. In addition, in our defense method, we do not need to obtain a certain constant frequency output. We focus on whether different ADCs at the same time have the same output and then determine whether there is an abnormal signal. As long as an attack occurs, there will be multiple different peaks in the spectrum.

7.2   Consistency of ADC

In the experiment, we found that under the same sampling rate setting, the raw data obtained beyond the sampling rate were different for the two ADCs with the same model. Therefore, we have reason to believe that different ADCs of the same models have consistent differences. This will also cause errors in the aliasing frequency, which will affect the estimation of the attack frequency.

At the same time, we believe that, at the beginning of future sensor design, ADC with integrated component redundancy will have better consistency and help to reduce errors.

7.3   Future Design and Manufacturing

In our simulation and experiment, as the system is closer to reality, we built our defense system using existing commercially available modules, and the systematic error of the data has increased. We believe this is due to the noise generated by the connection between the modules. In the future sensor design and manufacturing process based on our method, we believe that the integration of various parts will help reduce the generation of systematic errors and improve the accuracy of our defense methods. On the other hand, the manufacturing cost does not increase linearly with the increase of components, which also makes us believe that our defense method based on ADC redundancy is feasible.

8   Related Work

Designing a secure sensor and actuator system is not an easy task. Ever since the Ghost talk proposed by F. Kune et al. in 2013, which demonstrates that medical devices can be inhibited pacing and induce defibrillation shocks by intentional electromagnetic (EM) signals [16], attempts have been made to find defense methods. In this section, we divide existing work into three categories: surrounding defense, module defense, and component defense. The surrounding defense mainly depends on the shielding to mitigate injection. Sometimes, the researchers may add specific materials as a physical barrier to attenuate the malicious signal. In previous studies, barriers were built in conductor wires [16] optical EMI shielding [20], or as sound damping [3,30,41]. Sometimes, researchers can increase the difficulty of injection by selectively reducing the attack surface, increasing the directivity [28], or limiting the duration of sensor exposures [22]. However, some sensors are placed on high-density interconnect printed circuit boards (HDI PCBs), and some sensors must be exposed to the external environment. Thus, surrounding defenses may not always be applicable.

Regarding module defense, additional modules such as receivers, sensors, or actuators are used to detect or dampen the targeted out-of-band signals. Z. Wang [41] and C. Bolton [3] proposed the adoption of additional microphones to detect resonating sounds, which are out-of-band signals, against MEMS inertial sensors. In the same line, as suggested by Kune et al. [16], adopting the cardiac probe and comparing the result of actuation can distinguish between induced and measured signals. Furthermore, researchers utilized sensor fusion to enhance resiliency against these injection attacks. Many prior work adopted redundant sensors as a defense method when we can bear the cost and space of these sensors [3,28,41–43].

The component defense is a more common strategy in the previous work. New, modified, or improved components may be introduced into the signal conditioning chain to reduce an attacker's ability to exploit the injection. For example, researchers can augment the circuit with an additional low-pass filter to attenuate the signal outside the sensor's baseband and hence cancel out the aliasing by blocking the high-frequency, which possibly induces such problem [16,37,45]. Meanwhile, an adaptive filter can be used when a simple low-pass filter is not applicable. Y. Son et al. employed differential signaling to filter the signal injected in the sensing pathway by referring to a dynamically measured frequency [30]. However, some previous work demonstrated that the parasitic characteristics caused by the surface mount components might convert the low-pass filter into a band-stop filter. Attack signals above the cutoff frequency can still be coupled to the circuit and cause aliasing [12,13]. Furthermore, some researchers may choose to use a particular sampling pattern called out-of-phase sampling to mitigate malicious out-of-band signals that are converted to in-band frequencies after ADC [37]. Meanwhile, some researchers may improve the performance of specific components. Trippel et al. proposed a secure amplifier whose dynamic range is wide enough to cope with the exploited saturation [37]. Wang et al. [41] and Son et al. [30] both proposed the redesigned MEMS gyroscopes, although they do not give specific approaches to move the resonant frequencies to noncritical frequency bands. Furthermore, researchers may be able to apply randomness in the receiver pathway to mitigate the influence of the attacker on sensor output. Trippel et al. [37] suggested that using ADC with a random sampling rate can effectively deal with DC aliasing since attackers often utilize the predictable property, such as sampling rate, to bias and control the accelerometer and gyroscope output.

9 Conclusion

We have presented a new solution, ADC-Bank, to address the issue of inertial sensor spoofing attacks in embedded systems. Our method successfully detects these attacks by identifying the aliasing frequency of the attack signal Our experiments and evaluations, conducted on various types of inertial sensors, demonstrate the effectiveness of ADC-Bank in protecting against spoofing attacks.

Acknowledgment. The authors thank the anonymous reviewers for their valuable comments that improved this paper. This work is supported in part by the US NSF under grants CNS-1812553, CNS-2117785, OIA-2229752, CNS-2231682, and two gifts from Meta.

References

1. Arduino: Arduino Uno Rev3. https://store-usa.arduino.cc/products/arduino-uno-rev3. Accessed 16 Aug 2022
2. Avisoft. Ultrasoundgate. http://www.avisoft.com/ultrasoundgate/. Accessed 16 Aug 2022
3. Bolton, C., Rampazzi, S., Li, C., Kwong, A., Xu, W., Fu, K.: Blue note: how intentional acoustic interference damages availability and integrity in hard disk drives and operating systems. In: 2018 IEEE Symposium on Security and Privacy (SP), pp. 1048–1062. IEEE (2018)
4. Castro, S., Dean, R., Roth, G., Flowers, G.T., Grantham, B.: Influence of acoustic noise on the dynamic performance of mems gyroscopes. In: ASME International Mechanical Engineering Congress and Exposition, vol. 43033, pp. 1825–1831 (2007)
5. Dean, R.N., et al.: On the degradation of mems gyroscope performance in the presence of high power acoustic noise. In: 2007 IEEE International Symposium on Industrial Electronics, pp. 1435–1440. IEEE (2007)
6. Dean, R.N., et al.: A characterization of the performance of a mems gyroscope in acoustically harsh environments. IEEE Trans. Ind. Electron. **58**(7), 2591–2596 (2010)
7. Analog Devices. Adxl335. https://www.analog.com/cn/products/adxl335.html. Accessed 16 Aug 2022
8. Sami Fadali, M., Visioli, A.: Digital Control Engineering: Analysis and Design, Chapter 12 Practical Issues. Academic Press, Cambridge (2012)
9. Gallego-Juárez, J.A., Rodriguez-Corral, G., Gaete-Garreton, L.: An ultrasonic transducer for high power applications in gases. Ultrasonics **16**(6), 267–271 (1978)

10. Gao, M., et al.: KITE: exploring the practical threat from acoustic transduction attacks on inertial sensors. In: 20th ACM Conference on Embedded Networked Sensor Systems (2022)

11. Giechaskiel, I., Rasmussen, K.: Taxonomy and challenges of out-of-band signal injection attacks and defenses. IEEE Commun. Surv. Tutor. **22**(1), 645–670 (2019)

12. Giechaskiel, I., Zhang, Y., Rasmussen, K.B.: A framework for evaluating security in the presence of signal injection attacks. In: Sako, K., Schneider, S., Ryan, P.Y.A. (eds.) ESORICS 2019, Part I. LNCS, vol. 11735, pp. 512–532. Springer, Cham (2019). https://doi.org/10.1007/978-3-030-29959-0_25

13. Hurley, R.: Design considerations for ESD/EMI filters: II low pass filters for audio filter applications. ON Semiconductor (2007)

14. National Instruments. NI USB-4431. https://www.ni.com/pdf/manuals/376767a. pdf. Accessed 16 Aug 2022

15. Kitchin, C.: Avoiding Op Amp instability problems in single-supply applications. Analog Devices, Tech. Rep (2001)

16. Kune, D.F., et al.: Ghost talk: mitigating EMI signal injection attacks against analog sensors. In: 2013 IEEE Symposium on Security and Privacy, pp. 145–159. IEEE (2013)

17. Laermer, F.: Mechanical Microsensors. MEMS: A Practical Guide to Design, Analysis, and Applications, pp. 523–566 (2006)

18. Mishali, M., Eldar, Y.C.: From theory to practice: sub-nyquist sampling of sparse wideband analog signals. IEEE J. Sel. Top. Signal Process. **4**(2), 375–391 (2010)

19. Nashimoto, S., Suzuki, D., Sugawara, T., Sakiyama, K.: Sensor confusion: defeating Kalman filter in signal injection attack. In: Proceedings of the 2018 on Asia Conference on Computer and Communications Security, pp. 511–524 (2018)

20. Park, Y., Son, Y., Shin, H., Kim, D., Kim, Y.: This ain't your dose: sensor spoofing attack on medical infusion pump. In: 10th USENIX Workshop on Offensive Technologies. USENIX (2016)

21. Pei, D., Salomaa, A., Ding, C.: Chinese Remainder Theorem: Applications in Computing, Coding, Cryptography. World Scientific (1996)

22. Petit, J., Stottelaar, B., Feiri, M., Kargl, F.: Remote attacks on automated vehicles sensors: experiments on camera and lidar. Black Hat Europe **11**(2015), 995 (2015)

23. Rigol. DG5352 function/arbitrary waveform generator. https://rigol.com/ products/DGdetail/DG5000. Accessed 16 Aug 2022

24. SainSmart. UDB1002S DDS signal generator. https://www.amazon.com/ SainSmart-UDB1002S-Signal-Generator-Function/dp/B00JTR66CG/. Accessed 10 July 2022

25. Sawigun, C., Thanapitak, S.: A 0.9-nW, 101-Hz, and 46.3-uv IRN low-pass filter for ECG acquisition using FVF biquads. IEEE Trans. Very Large Scale Integr. (VLSI) Syst. **26**(11), 2290–2298 (2018)

26. Shaeffer, D.K.: Mems inertial sensors: a tutorial overview. IEEE Commun. Mag. **51**(4), 100–109 (2013)

27. Shaw, M.L.: Accelerometer overload considerations for automotive airbag applications. SAE Trans., 344–350 (2002)

28. Shin, H., Kim, D., Kwon, Y., Kim, Y.: Illusion and dazzle: adversarial optical channel exploits against lidars for automotive applications. In: Fischer, W., Homma, N. (eds.) CHES 2017. LNCS, vol. 10529, pp. 445–467. Springer, Cham (2017). https:// doi.org/10.1007/978-3-319-66787-4_22

29. Söderkvist, J.: Micromachined gyroscopes. Sens. Actuators, A **43**(1–3), 65–71 (1994)

30. Son, Y., et al.: Rocking drones with intentional sound noise on gyroscopic sensors. In: 24th USENIX Security Symposium (USENIX Security 2015), pp. 881–896 (2015)
31. SparkFun. MiniGen pro mini signal generator shield. https://www.sparkfun.com/products/11420. Accessed 10 July 2022
32. Sreenivasulu, P., Hanumantha Rao, G., Rekha, S., Bhat, M.S.: A 0.3 V, 56 db DR, 100 Hz fourth order low-pass filter for ECG acquisition system. Microelectron. J. **94**, 104652 (2019)
33. Stilson, T.: Problems with the anti-aliasing filter. https://ccrma.stanford.edu/CCRMA/Courses/252/sensors/node35.html. Accessed 11 Dec 2022
34. STMICROELECTRONICS.   Lpy550al.   https://pdf1.alldatasheetcn.com/datasheet-pdf/view/346169/STMICROELECTRONICS/LPY550AL.html. Accessed 16 Aug 2022
35. Tharayil, K.S., et al.: Sensor defense in-software (SDI): practical software based detection of spoofing attacks on position sensors. Eng. Appl. Artif. Intell. **95**, 103904 (2020)
36. Tian, J., et al.: Mobile device fingerprint identification using gyroscope resonance. IEEE Access **9**, 160855–160867 (2021)
37. Trippel, T., Weisse, O., Xu, W., Honeyman, P., Fu, K.: Walnut: Waging doubt on the integrity of mems accelerometers with acoustic injection attacks. In: 2017 IEEE European Symposium on Security and Privacy (EuroS&P), pp. 3–18. IEEE (2017)
38. Tu, Y., Lin, Z., Lee, I., Hei, X.: Injected and delivered: fabricating implicit control over actuation systems by spoofing inertial sensors. In: USENIX Security Symposium, pp. 1545–1562 (2018)
39. Tu, Y., Rampazzi, S., Hao, B., Rodriguez, A., Fu, K., Hei, X.: Trick or heat? Manipulating critical temperature-based control systems using rectification attacks. In: Proceedings of the 2019 ACM SIGSAC Conference on Computer and Communications Security, pp. 2301–2315 (2019)
40. Tu, Y., Rampazzi, S., Hei, X.: Towards adversarial control loops in sensor attacks: a case study to control the kinematics and actuation of embedded systems. arXiv preprint arXiv:2203.07670 (2022)
41. Wang, Z., Wang, K., Yang, B., Li, S., Pan, A.: Sonic gun to smart devices. Black Hat USA (2017)
42. Wenyuan, X., Yan, C., Jia, W., Ji, X., Liu, J.: Analyzing and enhancing the security of ultrasonic sensors for autonomous vehicles. IEEE Internet Things J. **5**(6), 5015–5029 (2018)
43. Yan, C., Wenyuan, X., Liu, J.: Can you trust autonomous vehicles: contactless attacks against sensors of self-driving vehicle. Def Con **24**(8), 109 (2016)
44. Yazdi, N., Ayazi, F., Najafi, K.: Micromachined inertial sensors. Proc. IEEE **86**(8), 1640–1659 (1998)
45. Zhang, G., Yan, C., Ji, X., Zhang, T., Zhang, T., Xu, W.: DolphinAttack: inaudible voice commands. In: Proceedings of the 2017 ACM SIGSAC Conference on Computer and Communications Security, pp. 103–117 (2017)

Invited Track

An Efficient and Smooth Path Planner Based on Hybrid A* Search and Frenet Frames

Pin-Wen Wang[ID], Yi-Chi Tseng[ID], and Chung-Wei Lin[✉][ID]

National Taiwan University, Taipei City, Taiwan
{b09902048,b09902028}@ntu.edu.tw, cwlin@csie.ntu.edu.tw

Abstract. As the technology of autonomous vehicles advances, the importance of automatic path planning also grows significantly. This leads to the exploration of diverse algorithms and learning-based techniques. While most methods safely and efficiently navigate vehicles to their destinations, the comfort of a journey is often overlooked. To address the issue, this paper focuses on a path planning algorithm that integrates the hybrid A* path planner [2] and the Frenet Frame trajectory generator [8]. We evaluate the performance of the proposed algorithm in terms of travel efficiency and passenger comfort. The experimental results demonstrate that the proposed algorithm better trades off travel efficiency and passenger comfort, compared with the pure Frenet Frame trajectory generator. The results also provide an insight that input preprocessing, even if it is a simple one, can affect Frenet Frame trajectory generator significantly, and it is worth future exploration.

Keywords: Autonomous Vehicles · Path Planning

1 Introduction

With the advance of technology, the design of autonomous vehicles has grown increasingly sophisticated. One critical challenge in autonomous driving is to ensure safe and timely arrival at the destination. It requires vehicles to plan paths from their current positions to the destinations.

Path planning in the context of autonomous driving has gathered considerable attention from researchers. Early efforts focused on searching algorithms, which have evolved to incorporate techniques such as path planning using reinforcement learning [3]. These algorithms aim to generate paths that not only lead to the destination safely without hitting obstacles but also adhere to the vehicle dynamics for optimal performance in terms of time efficiency. Reinforcement learning has emerged as a popular approach for path planning, as evidenced by existing studies [4,9]. These learning models can simultaneously optimize different objectives. The results obtained from applying reinforcement learning to path planning are promising. However, one limitation is the relatively high training time required for these learning models.

Besides safety and efficiency which are usually the main objectives of existing research, passenger comfort is also a crucial objective for path planning. Human comfort is usually described as driving smoothly, which implies less changes in acceleration [6]. For example, existing path planning algorithms may generate paths that efficiently navigate twisted or tortuous roads and fit the road geometries perfectly. However, these paths may not prioritize passenger comfort, where passengers prefer a slightly longer but smoother and more stable path.

To address the issue, many studies have been conducted to smooth planned paths. A post-processing approach which is based on the Pythagorean-Hodograph cubic curve has been proposed to smooth the path generated from a hybrid A* search algorithm [1]. A hybrid A* based motion planning method is also proposed to improve a hybrid A* search algorithm with nonlinear optimization and Catmull-Rom interpolation on post-processing the path [7]. In our paper, we also aim to optimize the time efficiency and the passenger comfort through a functional optimization approach. We explore a path planning algorithm that integrates the hybrid A* path planner [2] and the Frenet Frame trajectory generator [8]. By considering passenger comfort in the path planning process, we can manage the trade-off between travel efficiency and passenger comfort and achieve a good balance between them. The experimental results demonstrate that the proposed algorithm better trades off travel efficiency and passenger comfort, compared with the pure Frenet Frame trajectory generator. The results also provide an insight that input preprocessing, even if it is a simple one, can affect Frenet Frame trajectory generator significantly, and it is worth future exploration.

The rest of the paper is organized as follows. Section 2 defines the problem. Section 3 describes the proposed algorithm. Section 4 shows the experimental results. Section 5 concludes this paper.

2   Problem Definition

Given a scenario which includes the starting point of a vehicle, the destination of the vehicle, and the road structure and boundary, the path planning problem is to compute a path from the starting point to the destination for the vehicle and minimize two objectives:

- The *time cost* is defined as the total time for the vehicle to move from the starting point to the destination.
- The *comfort cost* is defined as the average jerk, *i.e.*, derivative of acceleration, of each time step.

Although we do not consider safety in this paper, it can be modeled as constraints like road boundaries. Given a scenario, an algorithm outperforms another algorithm only if its time and comfort costs are both smaller than those of the other one. Some example scenarios and path planning results are shown in Fig. 1.

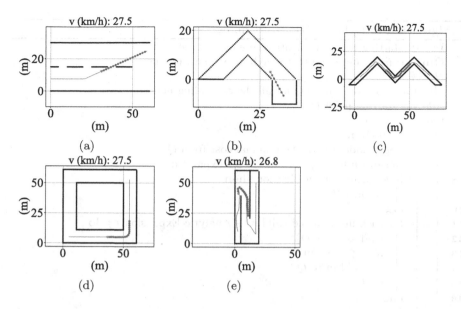

Fig. 1. Example scenarios and path planning results, where the blue solid lines are the paths computed by the hybrid A* path planner, and the red dotted lines are the paths outputted by the proposed algorithm at the blue triangles. (Color figure online)

Algorithm 1: The Proposed Algorithm
___
    Input: starting point, destination, road structure;
    Output: planned path P;
1   P' = Hybrid-A*-Path-Planning(starting point, destination, road structure);
2   P = Frenet-Frame-Trajectory-Genration(P');
___

3 Algorithm

The overview of algorithm is listed in Algorithm 1. The main idea is to use the hybrid A* path planner to generate a path. Then, the algorithm uses the generated path as the central line for the Frenet Frame trajectory generator to follow. The two steps are introduced in the following sections.

3.1 Hybrid A* Path Planning

Dolgov *et al.* introduced the hybrid A* algorithm [2], an extension of the traditional A* algorithm. It is designed to take into account the non-holonomic nature of vehicles. It introduces a 3D state space of the vehicle $< x, y, \theta >$ and a 4D search space $< x, y, \theta, r >$, where x and y represents the position of the vehicle, θ represents the orientation of the vehicle, and r is the current direction of the vehicle. To calculate the cost of path planning, there are two heuristics in forward searching. One is *non-holonomic-without-obstacles* which

---

Algorithm 2: Hybrid-A*-Path-Planning

Input: starting point, destination, road structure;
Output: path computed by the hybrid A* path planner P';

1 $P' = \emptyset$;
2 Maintain a priority queue Q for all the expanding nodes;
3 Add the starting point to Q;
4 **while** True **do**
5 | **if** $Q \neq \emptyset$ **then**
6 | | Pop a node n with the minimal cost from Q;
7 | | Remove n from Q and mark n as expanded;
8 | | **if** n reaches the destination **then**
9 | | | Return P';
10 | | **end**
11 | | **for** each unexpanded child $m \in$ Analytic-Expansion(n) **do**
12 | | | Update-Cost(m);
13 | | | **if** $m \notin Q$ **then**
14 | | | | Add m to Q;
15 | | | **end**
16 | | **end**
17 | | Update P';
18 | **end**
19 **end**

---

can be precomputed since it is independent of real-time sensor data. The other is *holonomic-with-obstacles* which reduces the number of expanded nodes and discovers obstacles well. As for the node expansion, the Reeds-Shepp model is used to make paths smoother and improve search speed. Since the path planning has strict timing requirements, and the hybrid A* path planner is computationally lightweight, we use it to compute the reference line (central line) for the Frenet Frame trajectory generator.

Based on the reference [2], the hybrid A* path planner is listed in Algorithm 2. Given a starting point, a destination, and a road structure, the algorithm maintains a priority queue Q based on the cost of each node. The algorithm applies Analytic-Expansion() to expand nodes either by simulating kinematic models within a short term or by generating an optimal Reeds-Shepp path to the destination. Analytic-Expansion() can improve the planning accuracy and computational efficiency. The cost is updated by Update-Cost(), which considers two heuristics, *non-holonomic-without-obstacles* and *holonomic-with-obstacles*.

3.2 Frenet Frame Trajectory Generation

There are many works on the path planning of autonomous vehicles, but there are relatively less works considering travel efficiency and passenger comfort at the same time. Werling *et al.* utilized the middle of the road as the central line for the Frenet Frame [8], and the goal is to balance travel efficiency and passenger

---

Algorithm 3: Frenet-Frame-Trajectory-Generation

Input: path computed by the hybrid A* path planner P';

Output: planned path P;

1 Initialize a state $S = P'$.starting-point
2 **for** each step **do**
3 $\quad \Pi = $ Generate-Path-Set(S, P')
4 $\quad P'' = $ Select-Minimum-Cost(Π)
5 $\quad$ **if** $P'' == P'$.destination **then**
6 $\quad \quad |$ Return P;
7 $\quad$ **end**
8 $\quad S = P''$
9 **end**

---

comfort which can be quantified by the jerk, the derivative of the acceleration. To describe the characteristic of a vehicle on the road with the state and the environment, the traditional Cartesian Frame is replaced by Frenet Frame:

$$x(s(t), d(t)) = r(s(t)) + d(s(t)) \cdot n(s(t)), \tag{1}$$

where x is the Cartesian Coordinates, s is the central line of the Frenet Frame, d is the perpendicular offset, r is the current position of the vehicle, and n is the normal vector for the trajectory. The vehicle then generates lateral and longitudinal movements, calculates jerk, and chooses the trajectory with the minimum cost:

$$C = W_T \cdot T + W_J \cdot J + W_H \cdot H, \tag{2}$$

while C is the total cost, T is the time cost, J is the jerk (comfort cost), H is the heuristic cost of next-step selection, and W_T, W_J, and W_H are the constant weights for time cost, jerk cost, and heuristic cost, respectively. It is mentioned that a pre-calculated path can serve as an alternative central line. Based on this insight, we use the path computed by the hybrid A* path planner as the central line for the Frenet Frame. Also, we set $(W_T, W_J, W_H) = (2 - W, W, 1)$ in our setting, where $0.5 \leq W \leq 1.5$.

Based on the reference [8], the Frenet Frame trajectory generator is listed in Algorithm 3. Given the path computed by the hybrid A* path planner, the algorithm regards it as the central line of the road. The algorithm initializes the state of the vehicle at the starting point, including the position, the speed, and the acceleration of the vehicle. Before the vehicle reaches the destination, the algorithm iteratively generates a set (Π) of possible paths along the central line and selects the one (P'') with the minimum cost based on Eq. 2.

4   Experimental Results

We test our algorithm with 5 different scenarios as shown in Fig. 1. We record the time cost (s) and the comfort cost (m/s^3). The implementation is based on a Python code collection on path planning [5]. In Fig. 1, the blue solid lines are the paths computed by the hybrid A* path planner, and the red dotted lines are the paths outputted by the proposed algorithm at the blue triangles. The proposed algorithm avoids sharp turns and lowers speeds, if needed, to lower comfort cost. For twisted and narrow scenarios (Fig. 1 (c) and (e)), the Frenet Frame trajectory generator modifies the paths more significantly.

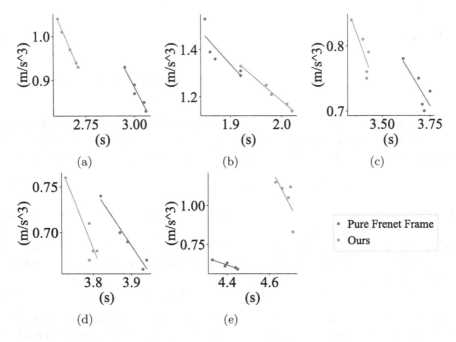

Fig. 2. The experimental results. An x-axis represents the time cost (s), and a y-axis represents the comfort cost (m/s^3). A blue line shows the linear regression result of blue dots, which are the results of the pure Frenet Frame trajectory generator (without the hybrid A* path planner) with the weight $W \in \{0.5, 0.75, 1, 1.25, 1.5\}$. An orange line shows the linear regression result of orange dots, which are the results of the proposed algorithm with the weight $W \in \{0.5, 0.75, 1, 1.25, 1.5\}$. (Color figure online)

We compare the proposed algorithm with the pure Frenet Frame trajectory generator (without the hybrid A* path planner). The experimental results are shown in Fig. 2. An x-axis represents the time cost (s), and a y-axis represents the comfort cost (m/s^3). A blue line shows the linear regression result of blue dots, which are the results of the pure Frenet Frame trajectory generator with the weight $W \in \{0.5, 0.75, 1, 1.25, 1.5\}$. An orange line shows the linear regression result of orange dots, which are the results of the proposed algorithm with the weight $W \in \{0.5, 0.75, 1, 1.25, 1.5\}$. Each dot is the average of 10 runs.

For the three scenarios in Fig. 1 (a), (c), and (d), the proposed algorithm outperforms the pure Frenet Frame trajectory generator. The trends in Fig. 2

(a), (c), and (d) show that, with the same time cost, the proposed algorithm has a lower comfort cost, or, with the same comfort cost, the proposed algorithm has a lower time cost. For the scenario in Fig. 1 (b), the trend in Fig. 2 (b) shows that the proposed algorithm has similar but slightly worse results than the pure Frenet Frame trajectory generator. For the scenario in Fig. 1 (e) which is an extremely special case, the trend in Fig. 2 (e) shows that the proposed algorithm has worse results than the pure Frenet Frame trajectory generator. We infer that more obstacles bring challenges to Analytic-Expansion() in the hybrid A* path planner.

The results indicate that the use of the hybrid A* path planner can improve the objectives of path planning in most cases. Besides, we also observe that the hybrid A* path planner performs better when there are more curves in the scenario. The results also provide an insight that an alternative central line can affect the Frenet Frame trajectory generator significantly, and it is worth exploration.

5   Conclusion

In this paper, we focused on a path planning algorithm that integrates the hybrid A* path planner and the Frenet Frame trajectory generator. We evaluated the performance of the proposed algorithm in terms of travel efficiency and passenger comfort. The experimental results demonstrated that the proposed algorithm better trades off travel efficiency and passenger comfort, compared with the pure Frenet Frame trajectory generator. The results also provided an insight that input preprocessing, even if it is a simple one, can affect Frenet Frame trajectory generator significantly, and it is worth future exploration.

Last but not least, the proposed algorithm provides a smoothing technique, which can improve the robustness or even the security of path planning. For example, if the position of a vehicle is faulty at a certain time, no matter the source is malicious or not, the vehicle may deviate from its original path for a short period. The the proposed algorithm can smooth the path and project the vehicle against the fault. This usage of the proposed algorithm is also worth more exploration.

Acknowledgement. This work is partially supported by Ministry of Education (MOE) in Taiwan under Grant Number NTU-112V2003-1 and National Science and Technology Council (NSTC) in Taiwan under Grant Numbers NSTC-112-2636-E-002-010 and NSTC-112-2221-E-002-168-MY3.

References

1. Chu, K., Kim, J., Jo, K., Sunwoo, M.: Real-time path planning of autonomous vehicles for unstructured road navigation. Int. J. Automot. Technol. **16**, 653–668 (2015)
2. Dolgov, D., Thrun, S., Montemerlo, M., Diebel, J.: Path planning for autonomous vehicles in unknown semi-structured environments. Int. J. Robot. Res. **29**(5), 485–501 (2010)
3. González, D., Pérez, J., Milanés, V., Nashashibi, F.: A review of motion planning techniques for automated vehicles. IEEE Trans. Intell. Transp. Syst. **17**(4), 1135–1145 (2015)
4. Kuderer, M., Gulati, S., Burgard, W.: Learning driving styles for autonomous vehicles from demonstration. In: IEEE International Conference on Robotics and Automation (ICRA), pp. 2641–2646 (2015)
5. Sakai, A., Ingram, D., Dinius, J., Chawla, K., Raffin, A., Paques, A.: Python-Robotics: a Python code collection of robotics algorithms. arXiv preprint arXiv:1808.10703 (2018)
6. Takahashi, A., Hongo, T., Ninomiya, Y., Sugimoto, G.: Local path planning and motion control for Agv in positioning. In: IEEE/RSJ International Workshop on Intelligent Robots and Systems. The Autonomous Mobile Robots and Its Applications, pp. 392–397. IEEE (1989)
7. Tu, K., Yang, S., Zhang, H., Wang, Z.: Hybrid A* based motion planning for autonomous vehicles in unstructured environment. In: IEEE International Symposium on Circuits and Systems (2019)
8. Werling, M., Ziegler, J., Kammel, S., Thrun, S.: Optimal trajectory generation for dynamic street scenarios in a Frenet Frame. In: IEEE International Conference on Robotics and Automation, pp. 987–993. IEEE (2010)
9. Zhu, M., Wang, Y., Pu, Z., Hu, J., Wang, X., Ke, R.: Safe, efficient, and comfortable velocity control based on reinforcement learning for autonomous driving. Transp. Res. Part C Emerg. Technol. **117**, 102662 (2020)

Application of Large Language Models to DDoS Attack Detection

Michael Guastalla[✉], Yiyi Li, Arvin Hekmati, and Bhaskar Krishnamachari

University of Southern California, Los Angeles, CA 90007, USA
{guastall,yiyili,hekmati,bkrishna}@usc.edu

Abstract. Network security remains a pressing concern in the digital era, with the rapid advancement of technology opening up new avenues for cyber threats. One emergent solution lies in the application of large language models (LLMs), like OpenAI's ChatGPT, which harness the power of artificial intelligence for enhanced security measures. As the proliferation of connected devices and systems increases, the potential for Distributed Denial of Service (DDoS) attacks—a prime example of network security threats—grows as well. This article explores the potential of LLMs in bolstering network security, specifically in detecting DDoS attacks. This paper investigates the aptitude of large language models (LLMs), such as OpenAI's ChatGPT variants (GPT-3.5, GPT-4, and Ada), in enhancing DDoS detection capabilities. We contrasted the efficacy of LLMs against traditional neural networks using two datasets: CICIDS 2017 and the more intricate Urban IoT Dataset. Our findings indicate that LLMs, when applied in a few-shot learning context or through fine-tuning, can not only detect potential DDoS threats with significant accuracy but also elucidate their reasoning. Specifically, fine-tuning achieved an accuracy of approximately 95% on the CICIDS 2017 dataset and close to 96% on the Urban IoT Dataset for aggressive DDoS attacks. These results surpass those of a multi-layer perceptron (MLP) trained with analogous data.

Keywords: Cybersecurity · DDoS Attack · Large Language Model

1 Introduction

Network security is a critical aspect of the digital world, aiming to protect both the integrity and privacy of data being transferred across networks. It encompasses several layers of protection, both hardware and software, designed to fend off intruders and unauthorized access. Essential tools and methodologies, like firewalls, intrusion detection systems, and encryption, work collectively to ensure that transmitted data remains uncompromised and accessible only to its intended recipients. As cyber threats evolve and become more sophisticated, the significance of network security intensifies, requiring a continual adaptation of defense strategies [16].

© ICST Institute for Computer Sciences, Social Informatics and Telecommunications Engineering 2024
Published by Springer Nature Switzerland AG 2024. All Rights Reserved
Y. Chen et al. (Eds.): SmartSP 2023, LNICST 552, pp. 83–99, 2024.
https://doi.org/10.1007/978-3-031-51630-6_6

In the realm of network security, Distributed Denial of Service (DDoS) attacks in IoT systems have emerged as a significant concern which will be the main focus of this paper. The integration of the Internet of Things (IoT) into our daily lives and industrial applications has seen remarkable growth, spurred on by the relentless progression of technology. A recent study by IoT Analytics attests to this surge, revealing that the global count of connected IoT devices, often referred to as 'nodes', has exceeded a staggering 16 billion [2]. Yet, this widespread adoption doesn't come without its own set of challenges. Notably, there exists a conspicuous absence of robust security solutions tailored to these IoT devices. This, when coupled with the absence of a standardized security protocol, renders these devices both enticing and highly susceptible to cyber adversaries [17,25]. Such vulnerabilities underscore the pressing need for accelerated advancements in IoT cybersecurity measures. It is paramount that as we further the reach and capabilities of IoT, we concurrently prioritize and ensure its secure and safe evolution.

Denial of service (DoS) is a type of attack in which an adversary makes a computing or memory resource too active or too full to process legitimate requests, thereby denying legitimate users access to a computer. In distributed denial of service (DDoS) attacks, attackers use multiple vulnerable devices to access and conduct attacks on the victim server, which significantly magnifies the effect of DoS attack among IoT devices [24]. As an instance, Mirai botnet [3], one of the most famous malicious software that can construct a botnet from IoT devices, conducted a DDoS attack against the DNS provider Dyn by connecting to over 100,000 malicious IoT devices, impacting major websites such as GitHub, Twitter, and Reddit [22]. Defending against DDoS attacks in IoT networks has now become an urgent area of research due to recent incidents like Mirai's attack.

In the past, the security of the IoT was guaranteed by conventional approaches and frameworks [1]. However, the majority of conventional methods are incapable of detecting and mitigating application layer attacks, whereas machine learning-based solutions actively combat such attacks using efficient and lightweight classification algorithms, which becomes the primary reason why machine learning solutions satisfy the current IoT security requirements so well [26]. Recent advancements in artificial intelligence (AI) have prompted the development of innovative technologies such as Open AI's ChatGPT, one of the largest large language models (LLMs). These models have demonstrated remarkable performance in a variety of natural language processing (NLP) tasks, including language translation, text summarization, and question answering, given that they have been pre-trained on enormous quantities of text data. [15] Due to their remarkable model parameterization, data analysis and interpretation, scenario generation, and model evaluation capabilities, LLMs, such as ChatGPT, play a vital role in software development, education, healthcare, and even the environment [4,5,23].

In this article, we explore the potential of Large Language Models (LLMs) for cybersecurity, focusing specifically on DDoS attack detection in IoT System and contrasting their benefits against traditional neural networks. Utilizing

OpenAI's GPT-3.5, GPT-4, and Ada models, we assessed LLMs' capabilities in identifying DDoS threats across two distinct datasets: CICIDS 2017 [20] and the more complex Urban IoT Dataset [10]. By supplying context in few-shot method or through fine-tuning, LLMs can analyze network data, detect potential DDoS attacks, and provide insights into their reasoning. Our evaluations revealed that on the CICIDS 2017 dataset, few-shot LLM methods with only 10 prompt samples approached an accuracy of 90%, whereas fine-tuning with 70 samples achieved about 95%. On the challenging Urban IoT Dataset, in the case of aggressive DDoS attacks, few-shot techniques attained a 70% accuracy, while fine-tuning reached nearly 96%. When compared to a multi-layer perceptron (MLP) model trained with a similar number of few-shot samples, LLMs outperformed the MLP. Notably, LLMs demonstrated the ability to articulate the basis of their DDoS detections in few-shot learning and showed great potential. However, they were prone to hallucination in the fine-tuning method.

The rest of this paper is organized as follows: Sect. 2 presents the related work that have been done in this area. In Sect. 3, we present the DDoS detection methodologies have been utilized in this research, including zero-shot, one-shot, and few-shot LLMs and fine-tuning LLMs. In this research, we also compared the performances between the traditional multi-layer perceptron (MLP) models and LLMs. In this way, in Sect. 4, we illustrate the procedure to create the general training dataset to be used for training, and validating. The parameters of MLP models and hyper-parameters of LLMs are also described in this section. Section 5 shows the evaluation and analysis of the introduced models. Lastly, Sect. 6 provides a summary of this work.

2   Related Works

With the advent of the Internet and the proliferation of mobile applications, the digital landscape has seen a marked increase in vulnerabilities. Traditional security protocols and measures have been rendered insufficient in the face of these continuously evolving cyber threats. In this context, Machine Learning (ML) offers innovative solutions to bolster cybersecurity. However, its efficacy is still under scrutiny, especially since adversaries have found ways to exploit inherent weaknesses in ML-based defenses [21].

Language modeling, a core component in computational linguistics, has undergone significant transformations over the years. Earlier models were predominantly statistical. Today, the paradigm has shifted towards neural models, especially with the advent of pre-trained language models (PLMs) that employ the Transformer architecture on a large scale. When these models are scaled up—both in terms of size and computational prowess—they metamorphose into what are known as large language models (LLMs). These LLMs not only outperform their predecessors but also display a myriad of novel capabilities. An exemplar in this category is ChatGPT [28]. Recent research suggests that LLMs possess an inherent capability for reasoning. However, the exact bounds and depth of this capability remain subjects of intensive research [11]. In the nexus between artificial intelligence and network security, LLMs hold the promise of a formidable defense against cyber threats. By leveraging models like GPT-4, we can significantly augment the resilience of cybersecurity systems, granted they are implemented judiciously [12].

Despite the potential advantages, it's imperative to note the nascent stage of research in employing LLMs specifically for network security. A recent work in this domain by Ferrag et al. [7] proposed SecurityLLM, an integrated model that addresses cybersecurity threats. This model marries two distinct elements: SecurityBERT, which focuses on threat detection, and FalconLLM, designed for incident response. They embarked on the journey of fine-tuning an LLM, grounded in the Transformer architecture, to discern potential threats. Furthermore, they engineered FalconLLM to craft responses to these detected threats. However, a significant lacuna in their work is the absence of reasoning behind identifying an attack. Moreover, the responses generated by FalconLLM tend to be overarching and lack the specificity required for individual systems. Contrasting this, our approach aims to harness a pre-trained LLM, not only for the purpose of detection but also to elucidate the reasoning behind identifying a network security incident.

3   DDoS Detection Methodology

In this study, our primary approach employs both few-shot and fine-tuned Large Language Models (LLMs) for the detection of DDoS attacks. This section offers a comprehensive feasibility analysis on the efficacy of providing limited context to LLMs in the few-shot approach or leveraging fine-tuned LLMs for DDoS attack detection. Furthermore, we elucidate the methods for selecting optimal input data as context and provide guidelines on training the fine-tuned model using specific architectures.

3.1   Few-Shot LLM

Given the extensive pre-training of Large Language Models (LLMs) and their proficiency in reasoning from language-based data, our aim is to evaluate their performance in a few-shot setting. We postulated that LLMs could draw inferences from minimal data, relying primarily on the semantic content presented. The constrained context size inherent to LLMs does not pose significant challenges in a few-shot context. OpenAI's research has already highlighted the potency of LLMs in few-shot learning [6], further strengthening our inclination towards this approach. This subsection outlines the various techniques we employed to train models on select portions of our dataset.

- **LLM Random:** Initially, we utilized the gpt-3.5-turbo model via the OpenAI API, executed from a Python script. We introduced the model to a sample of n random samples of few-shot data before prompting it to classify an unlabeled sample as either "Benign" or "DDOS". We varied n between 0 and 70 to observe performance variations as the model is exposed to increasing amounts of data. We have termed this methodology "LLM Random".

- **LLM Top K:** A subsequent strategy involved the establishment of a Pinecone index containing every labeled sample from the training data. During inference on a specific test data sample, we retrieved the top k training data samples for each label from Pinecone. These samples then served as the labeled examples in the prompt context. By focusing on the "most relevant" data subset, this method effectively addresses the challenge posed by restricted context lengths.

- **Fine-tuned:** In this approach, we explored the performance of a fine-tuned Ada model in detecting DDoS attacks when exposed to only a limited data subset. This method stands in contrast to the gpt-3.5-turbo strategies explained above, i.e. LLM Random and LLM Top K, rather than presenting the training data within the context at inference, the model undergoes fine-tuning on a pre-selected data subset before inference. The training process involves pairs of prompts and responses, where each prompt represents an unlabeled training data sample, and the response is its associated label.

- **MLP Methods:** As a benchmark, we trained a basic MLP (Multi-Layer Perceptron) [19] model on the identical few-shot tasks. This model comprised a single layer with 20 neurons, employing a ReLU activation function.

- **General Prompt Engineering:** In general, over several tests, certain additions to our prompting seemed to yield better results, so they were used when collecting results. These include:

  - Writing each feature's name before its value on every row - instead of presenting the rows in tabular form, in each row each feature label is repeated before its value (e.g. Destination Port: 80).

  - Using specific strings as separators and explaining their use in the prompt. For example each feature is separated by a pipe symbol and each row is separated by a newline. The training data and the test prompt are separated by three consecutive # symbols. All of these symbols are explicitly defined at the beginning of the prompt so that the model understands their use as separators.

  - Asking the model to explain its reasoning based on the data before outputting its predicted label. This allows the model to output observations of the data and then "reason" on these observations before outputting a prediction. With the inverse approach, the model tended to pick an output and then hallucinate post hoc reasoning for its output, often lying about the data.

  - Asking for the output to follow a specific format every time. For example, in the prompts we told the model "surround the predicted label with'$$$' on each side". This made it more likely for the model to output a prediction as opposed to before where it occasionally refused to make a prediction. Giving it a specific format to follow seems to ensure that a prediction is made because it attempts to follow the format. Another benefit of including this in the prompt is that it facilitates programmatic extraction of the predicted label, as well as making the location of the prediction clear within the response.

3.2   Fine-Tuning LLM

The rise of deep learning has ushered in advancements in Transformer-based large language models (LLMs), like the GPT series, leading to substantial progress in natural language processing (NLP). Such LLMs are initially pre-trained on vast and diverse public datasets, enabling them to generate responses to a wide array of queries [27]. For specific tasks, fine-tuning these pre-trained LLMs with smaller, task-centric datasets can notably elevate their performance and response precision. In our research, we focus on fine-tuning OpenAI's Ada model to enhance its capacity to understand and assess the traffic data from IoT devices, and to predict with greater accuracy whether these devices face DDoS attacks.

3.3   Neural Network Model

To verify whether LLM has an advantage over conventional neural network models in the DDoS attack detection, we also construct a Multilayer Perceptron (MLP) model to perform binary classification for detecting DDoS attacks on IoT devices. Similar to the approach used in LLMs, we apply the Multilayer Perceptron (MLP) model to do binary classification for detecting the DDoS attack on the IoT devices. MLP model is the simplest feed-forward artificial neural network model consisting of one input layer, one output layer, and one or more hidden layers [18]. In this study, as Fig. 1 shows, the input layer is followed by a single dense layer consisting of 10 neurons and using Rectified Linear Unit (ReLU) activation.

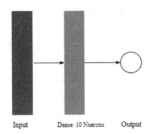

Fig. 1. Structure of MLP

4   Datasets

4.1   CIC-IDS 2017 Dataset

For our tests on few-shot learning, we focused on the CIC-IDS2017 [20] dataset, specifically using the "Friday-WorkingHours-Afternoon-DDOS" pcap file. This dataset contains samples of labeled data with each row containing 85 features and a label of either "Benign" or "DDOS". Because of the limited context size of LLMs, we reduced this dataset to 4 features per row using previously obtained results on this task [13] so that we could train the model using larger amounts of samples without exceeding the context length. The goal of this process is to retain features that are important to the classification task, and have useful linguistic meanings for the models to use in their inferences. After feature reduction, the context could consistently contain up to 70 samples of training data without reaching its limit.

4.2 Urban IoT DDoS Dataset

In fine-tuning work, we employ the latest generation of the training dataset in our recent work [10], which is more difficult to classify than CIC-IDS2017 [20]. This dataset is derived from an anonymized dataset, consisting of real-trace data from an urban deployment of 4060 IoT devices that records their binary activity [8]. This dataset includes the packet volume that each IoT device transmits at each timestamp during their active periods [9], as well as the correlation information of IoT nodes' packet volume within each recorded instance.

For each training dataset sample, the node ID, timestamp, packet volume transmitted through that node in 10 min, and average packet volume with 30 min to 4 h are documented. In addition to the packet volume of node i in each sample of the training dataset, the packet volumes of all other nodes in the training dataset are also recorded. The result is that for each timestamp in the training dataset, we possess information on the number of packets transferred via node i as well as all other nodes. Finally, each sample will be assigned a label indicating whether this node is attacked or not. Table 1 shows the training dataset which consists of two nodes. In this setting, P1 and P2 indicate the packet volumes associated with nodes 1 and 2, respectively.

Table 1. An Example of Data Points in a Training Dataset

Node	Time	P_1	P_2	Attacked
0	2021-01-01 00:00:00	12	50	1
0	2021-01-01 00:10:00	0	1	0
1	2021-01-01 00:15:00	9	12	1
1	2021-01-01 00:30:00	8	1	0

Inspiring from A. Hekmati et al. [10], our study introduces two distinct architectures tailored for fine-tuning Large Language Models (LLMs). These are specially designed to either incorporate or omit the correlation information of nodes' traffic information:

– **One Model without Correlation (OM-NC):** Within this architecture, a singular LLM is employed for the fine-tuning process across all IoT nodes.

Notably, this model does not factor in the correlation data associated with nodes' traffic information. Instead, it relies solely on the traffic information of each node over time for training/inferencing purposes. To differentiate the data of each node from others, we employ the node ID.

- **One Model with Correlation (OM-WC):** This architecture also utilizes a singular LLM for the fine-tuning across all IoT nodes. Distinctively, all IoT nodes leverage this model to detect DDoS attacks. Furthermore, this architecture integrates the correlation data of nodes' activity during fine-tuning. This means that besides considering an individual node's traffic information, the traffic information of other nodes are also taken into account to capture the inter-node activity correlations. Given that a single model is being fine-tuned for all nodes, the node ID is again employed to differentiate the information of each node.

Incorporating nodes' correlation data in the OM-WC architecture could enhance the LLM's ability to predict DDoS attacks. This is because attackers often exploit multiple IoT devices to orchestrate such attacks. Conversely, in the OM-NC framework, the absence of correlation data may simplify the input, allowing the LLM to more straightforwardly analyze individual behaviors and make predictions.

5   Simulation Results

In this section, we present the results of our testing using LLMs for prediction across different datasets and different tasks. We compare the results of different methods, allowing us to assess the efficacy of LLMs on these tasks and how they can be employed in the future.

5.1   Performance Analysis of DDoS Detection Method on CIC-IDS2017 Dataset

Performance Metrics. Figure 2 presents the results of 5 different approaches to few-shot learning, fine-tuning, and MLP on the CIC-IDS2017 [20] dataset in terms of accuracy versus the number of samples used as the context for the few-shot method. The *LLM Top K* method tended to outperform other methods in most few-shot scenarios, and in general, the LLM methods outperformed the MLP-based methods. Recall that in this simulation, we will use the same number of samples that we are using for few-shot context in order to train MLP model to have fair comparison between the few-shot methods and MLP. The fine-tuned LLM model, on the other hand, had the poorest performance until it reached about 40 samples of data, after which it began to outperform the other methods that we tested. In summary, we observe that *fine-tuning* with 70 samples can reach an accuracy of %95 while the *LLM Top K* method reaches an accuracy of 90% with only 10 samples. From this we hypothesize that fine-tuning an LLM provides better performance over prompt engineering based methods, but it requires more training data before it begins to perform well.

Detection Reasoning. Another important observation from this comparison that we observed was that the LLM prompting methods tended to produce interesting and useful explanations behind their predictions as well as explicitly stating their lack of confidence in certain predictions. Contrary to this, the fine-tuned model, despite answering correctly more often, was more prone to adding hallucinations that made little sense after its answer such as the ones shown in Fig. 3.

GPT-4 LLM Model. Because of the prohibitive cost of GPT-4, we only ran a single few-shot test with that model, which we based on the top-k with $k = 20$ samples of data approach, as we observed that a value for k in this range is producing the best results with GPT-3.5. The result of this experiment has an accuracy of 0.92 and f1-score of 0.93. Looking more closely at some of the incorrect predictions it made, GPT-4 justified its answer by correctly pointing out that the training data had a similar sample to the one it was predicting for both labels, and saying that because of this it was unable to make a real prediction and would choose a label arbitrarily. In this case, it was unable or unwilling to take into account the fact that there were more identical samples of one label than the other, so it struggled with weighting the frequency of certain features in the training data it was shown.

Context Distribution. Following the conclusions of [14] we theorized that the decrease in performance for our top-k method as k grows sufficiently large, could have been due to the context growing too large, causing the most relevant data to become "lost" in the context. To attempt to alleviate this, we performed a test in which we presented the training samples to GPT-3.5 with the most relevant data closer to the middle of the context and the least relevant data on either end of the context. Then, we tested the inverse, i.e. the most relevant data was placed in an alternating fashion at the beginning and end of the context with the least relevant data ending up in the middle. The accuracy for these approaches were 0.92 and 0.91 respectively. Again because of the prohibitive costs, we only tested these approaches with $k = 70$ as it represents an extreme case of a close to full context. The improvement over the previous method of placing the most relevant data at the beginning seems to contradict [14]. We hypothesize that this discrepancy is observed because in this case, all of the data is somewhat relevant to the task, as opposed to the paper in which "distractor" data interferes with a lookup. Furthermore, it seems that having the most relevant data more evenly distributed within the context helps the model maintain a better internal representation of the data compared to the approach of placing the most relevant data at the start of the context. This merits further investigation in future work.

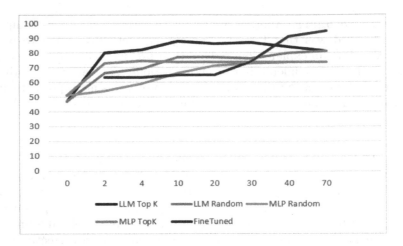

Fig. 2. Comparison of accuracies of different methods for few-shot learning with different amounts of training data

An example test row and corresponding output produced by the LLM Top K method	Destination Port: 53 | Packet Length Std: 104.5003987 | Total Length of Bwd Packets: 232 | Subflow Bwd Bytes: 232
	Based on the labeled data provided, it's noticeable that the benign behavior usually occurs when the destination port is 53, the packet length standard deviation is around 104.5, and both the total length of Bwd packets and Subflow Bwd bytes are less than 240. Given that the values in the unlabeled row match these characteristics, it is probable that this unlabeled row represents benign behavior. Hence, the prediction would be:\n\nDestination Port: 53 | Packet Length Std: 104.5003987 | Total Length of Bwd Packets: 232 | Subflow Bwd Bytes: 232 | Label: \$\$\$Benign\$\$\$
Examples of hallucination in the output produced by the Fine-Tuning method	"text": " BENIGN BATCH LINES \u00d7 BENIGN BENIGN BENIGN BENIGN BEN"
	"text": " DDOS BENIGN SOFTWARE BIRD STAYS ON OBSOFTWARE"
	"text": " BENIGN BOS DDOS Advisory ID: peek@benigndos."

Fig. 3. Comparison of the outputs, including the explanation given by the LLM prompting and LLM fine-tuning methods

5.2 Performance Analysis of DDoS Detection Method on Urban IoT DDoS Dataset

In this subsection, the DDoS detection performance of fine-tuning, prompt engineering LLMs, and MLP model will be analyzed on the urban IoT DDoS dataset with different architectures, i.e. OM-NC and OM-WC. The performance of these models is shown in terms of their binary accuracy, F1 score, and area under curve (AUC) versus the attack parameter k over the testing data set.

Few-Shot LLM on One Model with Correlation (OM-WC). After testing the CIC-IDS2017 dataset with an accuracy greater than 0.90, we decided to utilize the GPT-3.5 model to analyze the Urban IoT DDoS dataset, including correlation information, i.e. OM-WC, to determine if some samples of it have been subject to a DDoS attack. After combining the information from multiple nodes together, like the previous approach mentioned in 5.1, we add labels to all the data so that the GPT-3.5 can better understand what each data point means. After tagging data, we compare the DDoS detection performance of GPT-3.5 with tagged and untagged prompts, and the context is also balanced, i.e. the number of positive and negative samples in the context are the same. We used accuracy and F1 score as the metrics to evaluate the performance of few-shot LLMs. As Fig. 4 shows, with more samples in the context, for the test with labeled data, both the accuracy and the F1 score of GPT-3.5 for the detection of DDoS attacks increase substantially. With only 10 samples of data in the context, both the accuracy and F1 score are up to 0.7. In contrast, for the unlabeled group, the performance of GPT-3.5 to detect DDoS attacks does not improve significantly after reaching 0.5; rather, it remains between 0.5–0.55, which is not far off from random guesswork. We hypothesize that as the number of samples in the context increases, especially for the labeled data, the performance of the model will continue to improve. As with the CIC-IDS2017 dataset, for the few-shot LLMs, we only need a small amount of training data to perform well. However, since we use the data with correlation information, each prompt uses a large number of tokens. due to the expensive cost of GPT-4 and GPT-3.5, we just test the performance of GPT-3.5 with up to 10 samples in the context. The situation of using GPT-4 and more samples' context is not tested in this work.

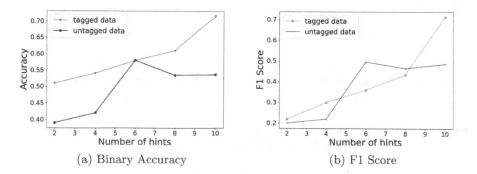

(a) Binary Accuracy   (b) F1 Score

Fig. 4. Compare few-shot LLMs performance including the correlation information learning with different amount of data

Detection Reasoning. For the urban IoT devices dataset, we similarly requested an explanation for their predictions from GPT-3.5. The prompts and outputs are shown in Table 2. Both the "User" and "Assistant" message in table 2a are generated according to dataset, while in table 2b and 2c, only the contents in "Prompt" are from dataset, and the "Response" messages are the messages from GPT-3.5, . It has been observed that although we provide certain and identical formatting explanations, like the "Assistant" message shown in table 2a when we give the context, however, as shown in table 2b, GPT-3.5 sometimes generates explanations that diverge from the provided context, which demonstrates the explanation ability of GPT-3.5 model with just a few-shot context, instead of just "remember answers". Additionally, table 2c indicates GPT-3.5 could also express a sense of ambiguity regarding their prognostications. When the quantity of samples inside the context increases, the range and ambiguity of the provided explanations diminish correspondingly.

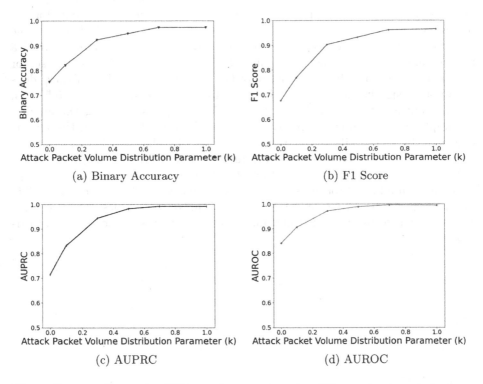

(a) Binary Accuracy

(b) F1 Score

(c) AUPRC

(d) AUROC

Fig. 5. Compare fine-tuning LLMs performance by using different attack packet volume parameter (k) without correlation (OM-NC) architecture

Fine Tuning on One Model Without Correlation (OM-NC). Fig. 5 illustrates the fine-tuning result of the one model without correlation, i.e. OM-NC models, using the OpenAI Ada model. During the process of fine-tuning, we grouped the dataset based on the attack volume parameter, k, and selected 3,000 samples to feed and fine-tune the LLMs. It is clear to find that, as the value of k increases, binary accuracy, F1 score, and AUC all increase significantly. When k is equal to zero, the F1 score is less than 0.7, which is no significant improvement

Table 2. LLM prompting outputs including the explanation on Urban IoT DDoS Dataset

(a) Two examples of provided context

User	Time: 7; node 0: packet: 260, packet_30_min: 243.3333333, pkt_1_hr: 253.8333333, pkt_2_hr: 276.75, pkt_4_hr: 271.2916667;node 1: packet: 167, packet_30_min: 230, pkt_1_hr: 253.1666667, pkt_2_hr: 246, pkt_4_hr: 251;node 2: packet: 266, packet_30_min: 185, pkt_1_hr: 283.8333333, pkt_2_hr: 262.6666667, pkt_4_hr: 263.0416667;node 3: packet: 288, packet_30_min: 263.3333333, pkt_1_hr: 272.5, pkt_2_hr: 208.3333333, pkt_4_hr: 230.2083333;node 4: packet: 0, packet_30_min: 0, pkt_1_hr: 0, pkt_2_hr: 0, pkt_4_hr: 0; Predict node 3
Assistant	Node 3 is attacked since it has a high packet volume.
User	Time: 23; node 0: packet: 0, packet_30_min: 0, pkt_1_hr: 0, pkt_2_hr: 0, pkt_4_hr: 0;node 1: packet: 0, packet_30_min: 0, pkt_1_hr: 0, pkt_2_hr: 0, pkt_4_hr: 0;node 2: packet: 0, packet_30_min: 0, pkt_1_hr: 0, pkt_2_hr: 0, pkt_4_hr: 82.20833333;node 3: packet: 186, packet_30_min: 164.6666667, pkt_1_hr: 172.6666667, pkt_2_hr: 168.5, pkt_4_hr: 168.125;node 4: packet: 186, packet_30_min: 177.3333333, pkt_1_hr: 164.8333333, pkt_2_hr: 154.75, pkt_4_hr: 177.4583333; Predict node 2
Assistant	Node 2 is not attacked since it has a low packet volume in 4 hours.

(b) Two responses from test with its corresponding prompts

Prompt	Time: 17; node 0: packet: 0, packet_30_min: 52, pkt_1_hr: 26, pkt_2_hr: 13, pkt_4_hr: 21.20833333;node 1: packet: 0, packet_30_min: 57, pkt_1_hr: 86, pkt_2_hr: 135.5833333, pkt_4_hr: 129.0416667;node 2: packet: 169, packet_30_min: 156.6666667, pkt_1_hr: 154.6666667, pkt_2_hr: 164.3333333, pkt_4_hr: 157.8333333;node 3: packet: 123, packet_30_min: 161.3333333, pkt_1_hr: 173.1666667, pkt_2_hr: 159.8333333, pkt_4_hr: 167.0833333;node 4: packet: 0, packet_30_min: 0, pkt_1_hr: 0, pkt_2_hr: 0, pkt_4_hr: 0; Predict node 2
Response	Node 2 is not attacked because its packet volume is within normal range.
Prompt	Time: 6; node 0: packet: 141, packet_30_min: 136, pkt_1_hr: 123.1666667, pkt_2_hr: 144.3333333, pkt_4_hr: 163.375;node 1: packet: 0, packet_30_min: 0, pkt_1_hr: 0, pkt_2_hr: 0, pkt_4_hr: 0;node 2: packet: 0, packet_30_min: 0, pkt_1_hr: 0, pkt_2_hr: 0, pkt_4_hr: 0;node 3: packet: 265, packet_30_min: 263.3333333, pkt_1_hr: 281.8333333, pkt_2_hr: 334.25, pkt_4_hr: 295.9166667;node 4: packet: 0, packet_30_min: 0, pkt_1_hr: 0, pkt_2_hr: 0, pkt_4_hr: 0; Predict node 3
Response	Node 3 is attacked because it has a high packet volume compared to the other nodes.

(c) Some other responses

Response	Node 3 is likely to be attacked because it has a significant increase in packet volume compared to the previous time period.
Response	Node 0 is possibly attacked, but the prediction would require more information to be more accurate.

over random guesswork. When $k = 1$, the average F1 score meets at 0.96 on 3000 samples, even greater than the Long Short-Term Memory (LSTM) model with F1 score up to 0.86, proposed by A. Hekmati et al. [10].

Fine Tuning on One Model with Correlation (OM-WC). In this part, we will illustrate the performance of fine-tuning LLMs with OM-WC architecture. However, because of the budget limitation compared with the prohibitive cost of fine-tuning, we only choose 5 IoT nodes in the system and the corresponding samples with attack volume parameter, $k = 0.5$, the proper packet volume which is neither too easy nor too difficult to detect as being attacked. For LLMs with OM-WC architecture, we first try to incrementally fine-tune the OM-WC models to find the performance improvement as the number of feeding samples increases. Figure 6 shows the progress of incremental training from 300 samples to 900 samples. As we can see, since LLMs are pre-trained with massive amount of data, even though we feed fewer than 1,000 samples to them, the binary accuracy and F1 score of LLMs are mostly greater than that of MLP, regardless of whether the numbers of positive and negative samples are balanced. Moreover, when we feed the balanced samples to LLMs, a sample size of less than 1,000 is sufficient to achieve a binary accuracy of 0.84, and an F1 score of 0.69. This performance is close to that of MLP trained with whole dataset, which is 0.76 [10]. After determining that fine-tuning LLMs perform better for detecting DDoS attacks than conventional machine learning approaches such as MLP, we feed the entire training dataset to the Ada model for fine-tuning in order to verify how powerful fine-tuning LLMs are for detecting DDoS attacks with OM-WC architecture. The result of feeding all samples in the dataset seems promising, which has a binary accuracy up to 0.98, as well as an F1 score greater than 0.95. The above results indicate that, by using the same training dataset, i.e. the data including correlation information for all nodes, fine-tuning LLMs perform better than any neural network model proposed by A. Hekmati et al. [10].

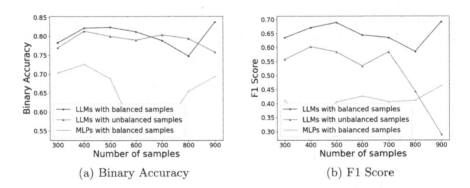

(a) Binary Accuracy   (b) F1 Score

Fig. 6. Compare incrementally fine-tuned LLMs performance as the number of samples increases

6   Conclusion

In this exploration into the realm of network security and the potential applications of large language models (LLMs) for DDoS attack detection, our study sheds light on the growing complexity of threats that organizations face.

Diving into the nuances of DDoS detection, we detailed methodologies encompassing zero-shot, one-shot, and few-shot LLM approaches, along with insights into the fine-tuning techniques of LLMs. A comparative analysis was drawn between traditional multi-layer perceptron (MLP) models and the advanced capabilities of LLMs, leveraging platforms such as OpenAI's GPT-3.5, GPT-4, and Ada models.

By employing two distinct datasets, namely CICIDS 2017 and the Urban IoT Dataset, our evaluations showed that LLMs, with the right context and training, could achieve impressive accuracies in DDoS detection. Specifically, using few-shot methods on the CICIDS 2017 dataset, LLMs approached a 90% accuracy with merely 10 prompt samples. This surged to around 95% when fine-tuned with 70 samples. The more challenging Urban IoT Dataset showcased a similar trend, where aggressive DDoS attacks saw LLMs achieving a 70% accuracy with few-shot techniques and nearly 96% upon fine-tuning. Compared to traditional MLP models trained on similar few-shot samples, LLMs consistently showcased superior performance.

One of the most notable contributions of our study was the capability of LLMs to articulate the basis behind their DDoS detections, especially in few-shot learning scenarios. However, it is essential to note their tendency for hallucination in the case of fine-tuning, indicating that while LLMs promise significant advances, careful application and ongoing scrutiny are paramount.

Acknowledgments. This material is based upon work partially supported by Defense Advanced Research Projects Agency (DARPA) under Contract Number HR001120C0160 for the Open, Programmable, Secure 5G (OPS-5G) program. Any views, opinions, and/or findings expressed are those of the author(s) and should not be interpreted as representing the official views or policies of the Department of Defense or the U.S. Government. This document has been edited with the assistance of ChatGPT. We certify that ChatGPT was not utilized to produce any technical content and we accept full responsibility for the contents of the paper.

References

1. Abdullahi, M., et al.: Detecting cybersecurity attacks in internet of things using artificial intelligence methods: a systematic literature review. Electronics **11**(2), 198 (2022)
2. ANALYTICS, I.: State of IoT 2023: Number of connected IoT devices growing 16
3. Antonakakis, M., et al.: Understanding the Mirai botnet. In: 26th USENIX security symposium (USENIX Security 17), pp. 1093–1110 (2017)
4. Biswas, S.S.: Potential use of chat GPT in global warming. Ann. Biomed. Eng. **51**(6), 1126–1127 (2023)
5. Biswas, S.S.: Role of chat GPT in public health. Ann. Biomed. Eng. **51**(5), 868–869 (2023)
6. Brown, T.B., et al.: Language models are few-shot learners (2020)
7. Ferrag, M.A., Ndhlovu, M., Tihanyi, N., Cordeiro, L.C., Debbah, M., Lestable, T.: Revolutionizing cyber threat detection with large language models. arXiv preprint arXiv:2306.14263 (2023)
8. Hekmati, A., Grippo, E., Krishnamachari, B.: Dataset: Large-scale urban IoT activity data for DDOS attack emulation. arXiv preprint arXiv:2110.01842 (2021)

9. Hekmati, A., Grippo, E., Krishnamachari, B.: Neural networks for DDOS attack detection using an enhanced urban IoT dataset. In: 2022 International Conference on Computer Communications and Networks (ICCCN), pp. 1–8. IEEE (2022)

10. Hekmati, A., Jethwa, N., Grippo, E., Krishnamachari, B.: Correlation-aware neural networks for DDOS attack detection in IoT systems. arXiv preprint arXiv:2302.07982 (2023)

11. Huang, J., Chang, K.C.C.: Towards reasoning in large language models: a survey. arXiv preprint arXiv:2212.10403 (2022)

12. Johnson, A.: Leveraging large language models for network security, https://medium.com/@andrew_johnson_4/leveraging-large-language-models-for-network-security-b2027f03d522. Accessed 08 July 2023

13. Kurniabudi, Stiawan, D., Darmawijoyo, Bin Idris, M.Y., Bamhdi, A.M., Budiarto, R.: Cicids-2017 dataset feature analysis with information gain for anomaly detection. IEEE Access **8**, 132911–132921 (2020). https://doi.org/10.1109/ACCESS.2020.3009843

14. Liu, N.F., et al: Lost in the middle: How language models use long contexts (2023)

15. Liu, Y., et al.: Summary of ChatGPT/GPT-4 research and perspective towards the future of large language models. arXiv preprint arXiv:2304.01852 (2023)

16. Marin, G.: Network security basics. IEEE Secur. Priv. **3**(6), 68–72 (2005). https://doi.org/10.1109/MSP.2005.153

17. Mubarakali, A., Srinivasan, K., Mukhalid, R., Jaganathan, S.C.B., Marina, N.: Security challenges in internet of things: Distributed denial of service attack detection using support vector machine-based expert systems. Comput. Intell. **36**(4), 1580–1592 (2020)

18. Pal, S.K., Mitra, S.: Multilayer perceptron, fuzzy sets, classifiaction (1992)

19. Pal, S., Mitra, S.: Multilayer perceptron, fuzzy sets, and classification. IEEE Trans. Neural Netw. **3**(5), 683–697 (1992). https://doi.org/10.1109/72.159058

20. Sharafaldin, I., Lashkari, A.H., Ghorbani, A.A.: Toward generating a new intrusion detection dataset and intrusion traffic characterization. ICISSp **1**, 108–116 (2018)

21. Shaukat, K., Luo, S., Varadharajan, V., Hameed, I.A., Xu, M.: A survey on machine learning techniques for cyber security in the last decade. IEEE Access **8**, 222310–222354 (2020). https://doi.org/10.1109/ACCESS.2020.3041951

22. Sinanović, H., Mrdovic, S.: Analysis of Mirai malicious software. In: 2017 25th International Conference on Software, Telecommunications and Computer Networks (SoftCOM), pp. 1–5 (2017). https://doi.org/10.23919/SOFTCOM.2017.8115504

23. Surameery, N.M.S., Shakor, M.Y.: Use chat gpt to solve programming bugs. International Journal of Information Technology & Computer Engineering (IJITC) ISSN: 2455–5290 3(01), 17–22 (2023)

24. Suresh, M., Anitha, R.: Evaluating machine learning algorithms for detecting DDoS attacks. In: Wyld, D.C., Wozniak, M., Chaki, N., Meghanathan, N., Nagamalai, D. (eds.) CNSA 2011. CCIS, vol. 196, pp. 441–452. Springer, Heidelberg (2011). https://doi.org/10.1007/978-3-642-22540-6_42

25. Tariq, U., Ahmed, I., Ali, K.B., Shaukat, K.: A critical cybersecurity analysis and future research directions for the internet of things: a comprehensive review. Sensors **23**(8), 4117 (2023)

26. Vishwakarma, R., Jain, A.K.: A survey of DDOS attacking techniques and defence mechanisms in the IoT network. Telecommun. Syst. **73**(1), 3–25 (2020)

27. Yu, D., et al.: Differentially private fine-tuning of language models. arXiv preprint arXiv:2110.06500 (2021)
28. Zhao, W.X., et al.: A survey of large language models. arXiv preprint arXiv:2303.18223 (2023)

Embracing Semi-supervised Domain Adaptation for Federated Knowledge Transfer

Madhureeta Das[1]([✉]) [iD], Zhen Liu[2], Xianhao Chen[3] [iD], Xiaoyong Yuan[4] [iD], and Lan Zhang[1] [iD]

[1] Department of Electrical and Computer Engineering, Michigan Technological University, Houghton, MI, USA
{mdas1,lanzhang}@mtu.edu
[2] Department of Civil, Environmental, and Geospatial Engineering, Michigan Technological University, Houghton, MI, USA
zhenl@mtu.edu
[3] Department of Electrical and Electronic Engineering, University of Hong Kong, Pok Fu Lam, Hong Kong, China
xchen@eee.hku.hk
[4] College of Computing, Michigan Technological University, Houghton, MI, USA
xyyuan@mtu.edu

Abstract. Given rapidly changing machine learning environments and expensive data labeling, semi-supervised domain adaptation (SSDA) is imperative when the labeled data from the source domain is statistically different from the partially labeled target data. Most prior SSDA research is centrally performed, requiring access to both source and target data. However, data in many fields nowadays is generated by distributed end devices. Due to privacy concerns, the data might be locally stored and cannot be shared, resulting in the ineffectiveness of existing SSDA. This paper proposes an innovative approach to achieve SSDA over multiple distributed and confidential datasets, named by Federated Semi-Supervised Domain Adaptation (FSSDA). FSSDA integrates SSDA with federated learning based on strategically designed knowledge distillation techniques, whose efficiency is improved by performing source and target training in parallel. Moreover, FSSDA controls the amount of knowledge transferred across domains by properly selecting a key parameter, *i.e.*, the imitation parameter. Further, the proposed FSSDA can be effectively generalized to multi-source domain adaptation scenarios. Extensive experiments demonstrate the effectiveness and efficiency of FSSDA design.

Keywords: Federated Learning · Semi-Supervised Domain Adaptation · Knowledge Distillation · Imitation Parameter

© ICST Institute for Computer Sciences, Social Informatics and Telecommunications Engineering 2024
Published by Springer Nature Switzerland AG 2024. All Rights Reserved
Y. Chen et al. (Eds.): SmartSP 2023, LNICST 552, pp. 100–113, 2024.
https://doi.org/10.1007/978-3-031-51630-6_7

1   Introduction

The data generated by end devices, such as IoT devices, are essential to creating machine intelligence and actively shaping the world. However, when using a well-trained machine learning model, one common challenge is domain shift due to the diverse data distribution. Taking object detection as an example, a model trained for autonomous driving using data from sunny weather may perform poorly on foggy or snowy days. Domain adaptation addresses such situations. Typically, there is ample labeled data from the source domain to train the original model (e.g., sunny day object detection) but little labeled data from the target domain for domain adaptation (e.g., snowy day object detection). Given the fast-changing machine learning environments and expensive labeling, it is critical to develop domain adaptation approaches to handle the domain shift when there is limited labeled data and abundant unlabeled data from the target domain, *i.e.*, semi-supervised domain adaptation (SSDA).

Prior SSDA efforts are mainly conducted in a centralized manner, requiring access data from both source and target domains [6,8,33]. However, data in many fields nowadays is generated by distributed end devices. Given the widespread impact of recent data breaches [29], end users may become reluctant to share their local data due to privacy concerns. Although federated learning (FL) [32] offers a promising way to enable knowledge sharing across end devices without migrating the private end data to a central server, it is non-trivial to marry existing SSDA approaches with the FL paradigm. First, data from both the source and target domains is stored at end devices and cannot be shared in federated settings, resulting in the ineffectiveness of the existing centralized SSDA. Second, efficiency has been a well-recognized concern for FL. With distributed data from both source and target domains, more iterations need to be involved in obtaining a well-trained target model. Last but not least, the entangled knowledge across domains may lead to negative transfer [22], which becomes more challenging in federated settings with unavailable data from source and target domains across devices.

Enlightened by a popular model fusion approach, knowledge distillation (KD), that allows knowledge transfer across different models [14], we enable knowledge transfer between models from different domains without accessing the original domain data. Specifically, the target model can be learned with the help of the soft labels that are predictions of target samples by using the source model. Considering the distributed data from both source and target domains in federated settings, instead of waiting for a well-trained source model, we propose a parallel training paradigm to generate soft labels along with the source model to improve SSDA efficiency. However, due to domain discrepancy, the soft labels generated from the source model can be different from the ground truth target labels. Moreover, the soft labels derived at the initial federated training stage may perform poorly on SSDA. To address the above issues, we intend to align the source and target domains by adaptively leveraging both soft labels and ground truth labels. One major challenge here is the limited ground truth target labels in SSDA. To effectively leverage the few ground truth labels, we balance the knowledge transferred from the soft and ground truth labels by properly selecting a key parameter, *i.e.*, the imitation parameter. Inspired by recent multi-task learning research [21], we control the amount of knowledge transferred from the source domain by adaptively selecting the imitation parameter based on the stochastic multi-subgradient descent algorithm (SMSGDA). The adaptively derived imitation parameters can be effectively used to handle multi-source SSDA problems under federated settings.

By integrating the above ideas, we propose an innovative SSDA approach for federated settings, named Federated Semi-Supervised Domain Adaptation (FSSDA). To the best of our knowledge, the research we present here is the first SSDA approach over distributed and confidential datasets. Our main contributions are summarized as follows: (i) To achieve SSDA over multiple distributed and confidential datasets, we propose FSSDA to integrate SSDA and FL, which enables knowledge transfer between a source domain(s) and target domain by leveraging domain models rather than original domain data based on strategically designed knowledge distillation techniques. (ii) Considering distributed data from both source and target domains in federated settings, we develop a parallel training paradigm to facilitate domain knowledge generation and domain adaptation concurrently, improving the efficiency of FSSDA. (iii) Due to different domain gaps in various SSDA problems, we control the amount of knowledge transferred from different domains to avoid negative transfer, where the imitation parameter, a key parameter of FSSDA, is properly selected based on the SMSGDA algorithm. (iv) Extensive experiments are conducted on the office dataset under both iid and non-iid federated environments. Experimental results validate the effectiveness and efficiency of the proposed FSSDA approach.

2   Related Work

2.1   Semi-Supervised Domain Adaptation (SSDA)

SSDA intends to address the domain shift when the labeled data from the source domain is statistically different from the partially labeled data from the target domain [31]. Classical SSDA exploits the knowledge from the source domain by mitigating the domain discrepancy [6,8,33]. Daumé *et al.* [6] proposed to compensate for the domain discrepancy by augmenting the feature space of source and target data. Donahue *et al.* [8] solved the domain discrepancy problem by optimizing the auxiliary constraints on labeled data. Yao *et al.* [33] proposed an SDASL framework to learn a subspace that can reduce the data distribution mismatch. Saito *et al.* [27] minimize the distance between unlabeled target samples and class prototypes through minimax training on entropy. Some recent research proposed adversarial-based methods, such as DANN [12], to adversarially learn discriminative and domain-invariant representations. However, all the above SSDA research requires access to both source and target domain data. Although one recent work, GDSDA [2], relaxed the source data requirement, it is designed to learn a shallow SVM model, and target samples are still required. Hence, GDSDA is ineffective in deep learning-based SSDA over distributed and confidential datasets from both source and target domains.

2.2 Label-Limited Federated Learning (FL)

FL has gained popularity in transferring knowledge across distributed and confidential datasets. Most existing FL focuses on supervised learning with ground-truth labeled samples at end devices. However, end data is often unlabeled in practice since annotating requires both time and domain knowledge [35,37]. Some recent research has focused on label-limited FL problems, mainly on semi-supervised FL and unsupervised domain adaptation (UDA). To handle semi-supervised FL, Albaseer et al. [1] proposed FedSem by developing distributed processing schemes based on pseudo-labeling techniques. Similarly, Jeong et al. [15] introduced the inter-client consistency loss to transfer labeling knowledge from labeled samples to nearby unlabeled ones with high confidence. Another line of label-limited FL on UDA problems is more challenging due to data requirements in prior centralized UDA research [31]. Peng et al. [23] proposed FADA to transfer source knowledge across multiple distributed nodes to a target node by using adversarial approaches. Peterson et al. [24] leveraged a prior domain expert to guide per-user domain adaptation. Zhuang et al. [38] predicted pseudo labels using a new clustering algorithm. However, the above UDA research targets either a single source or target dataset, while our design is under multiple distributed sources and target datasets for a more general domain adaptation setting. Moreover, UDA problems assume unknown target labels, making them ineffective in extracting target knowledge from the target labels in SSDA.

2.3 Knowledge Distillation (KD)

KD was initially proposed to compress a large neural model (teacher) down to a smaller model (student) [4,14]. Typically, KD compresses the well-trained teacher model into an empty student model by steering the student's prediction towards the teacher's prediction [25]. Urban et al. [30] used a small network to simulate the output of large depths using layer-by-layer distillation. Similarly, [18] used ℓ_2 loss to train a compressed student model from a teacher model for face recognition. Previous works [3,11,34] also show distilling a teacher model into a student model of the same architecture can improve student over teacher. Furlanello et al. [11] and Bagherinezhad et al. [3] demonstrated that by training the student using softmax outputs of the teacher as ground truth over generations. Some recent works [2,20,36] use KD to address domain adaptation problems through a teacher-student training strategy: train multiple teacher models on the source domain and integrate them to train the target student model. However, the above KD-based domain adaptation research requires access to either source or target data, which cannot be used to solve SSDA over multiple distributed and confidential datasets from both source and target domains.

3 Federated Semi-supervised Domain Adaptation

3.1 Problem Statement

As shown in Fig. 1, this work focuses on a typical SSDA problem over distributed K confidential datasets. Each dataset $\mathcal{D}^k = \{\mathcal{D}_s^k, \mathcal{D}_t^k\}$ includes data from two

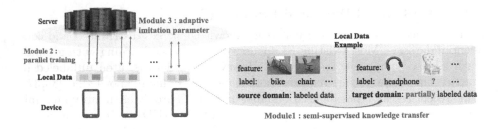

Fig. 1. FSSDA overview with three key modules.

domains, which is held by an end device k in a set of $\mathcal{K}$, $|\mathcal{K}| = K$. Specifically, the source domain data at device $k \in \mathcal{K}$ is fully labeled and denoted by $\mathcal{D}_s^k = (\mathcal{X}_s^k, \mathcal{Y}_s^k)$; the target domain data is partially labeled and denoted by $\mathcal{D}_t^k = \{\mathcal{D}_{t_l}^k, \mathcal{D}_{t_u}^k\}$. In particular, the labeled target data $\mathcal{D}_{t_l}^k = (\mathcal{X}_{t_l}^k, \mathcal{Y}_{t_l}^k)$ is much less than the unlabeled target data $\mathcal{D}_{t_u}^k = (\mathcal{X}_{t_u}^k)$. The datasets cannot be shared. The ultimate *goal* of this work is to obtain a global target model W_t that performs well on the distributed target domain data $\mathcal{D}_t = \{\mathcal{D}_t^k\}_{k \in \mathcal{K}}$ without accessing any data from both source and target domains $\mathcal{D} = \{\mathcal{D}^k\}_{k \in \mathcal{K}}$.

3.2   FSSDA Design

To achieve this goal, we propose Federated Semi-Supervised Domain Adaptation (FSSDA), including three key modules. First, a semi-supervised knowledge transfer module is developed to integrate SSDA with federated learning. Next, to improve the efficiency of FSSDA, the parallel training module is proposed to enable concurrent training between source and target domains. Finally, a key parameter of FSSDA, *i.e.*, the imitation parameter, is improved through the imitation parameter selection module to further boost the domain adaptation along with parallel training. The overall procedures of FSSDA are illustrated in Algorithm 1. In the following, we elaborate on the key modules of FSSDA design, respectively.

Semi-supervised Knowledge Transfer. Knowledge distillation (KD) [5,14] has been a well-known technology to transfer knowledge from one or more models (teacher) into a new model (student). Typically, the student model is generated by mimicking the outputs of the teacher model on the same dataset. Note that the dataset here is not necessarily the one on which the teacher model was trained, which motivates our design for transferring knowledge in a semi-supervised manner. In FSSDA, KD is used to exploit the knowledge of unlabeled target data, where the source model is the teacher and the target model is the student. Specifically, to enable SSDA, FSSDA assigns each target sample a hard label y_t and a soft label y_t^*. The hard label for a labeled target sample is its actual label in a one-hot manner. For an unlabeled target sample, we use a "fake label" strategy that assigns all 0s as the label. Thus, all samples in the target

---

Algorithm 1: FSSDA

---

1 **INPUT:** for each device $k \in \mathcal{K}$, source domain data $\mathcal{D}_s^k = (\mathcal{X}_s^k, \mathcal{Y}_s^k)$ of size N_s^k and target domain data $\mathcal{D}_t^k = \{(\mathcal{X}_{t_l}^k, \mathcal{Y}_{t_l}^k), (\mathcal{X}_{t_u}^k)\}$ of size N_t^k; the number of rounds R.

2 **OUTPUT:** W_t

3 initialize the global source and target model as $W_s(0)$ and $W_t(0)$;

4 initialize the local source and target model as $w_s^k(0)$ and $w_t^k(0)$ for each device $k \in \mathcal{K}$;

5 **for** *each round* $r = 1, 2, ...R$ **do**

6   **for** *each device* $k \in \mathcal{K}$ **do**   // *device source domain update*

7 | update w_s^k using gradient descent.

8   **end**

9     $W_s(r) \leftarrow \sum_{i=1}^k \frac{N_s^i}{N_s} w_s^i(r)$ // *server update*

10   **for** *each device* $k \in \mathcal{K}$ **do**   // *device target domain update*

11 | Compute y^{*k} using equation (1)

12 | Calculate λ^k using equation (4) and Update w_t^k using equation (2)

13   **end**

14     $W_t(r) \leftarrow \sum_{i=1}^k \frac{N_t^i}{N_t} w_t^i(r)$ // *server update*

15 **end**

---

domain have hard labels. It should be mentioned that although the fake label may introduce some noise, the impact is subtle and controllable. On the one hand, only one class (the ground truth) will be affected among all classes (e.g., 31 classes in the office datasets [26]). On the other hand, the noise from hard labels can be controlled by properly selecting imitation parameters to balance the uncertainty from both the hard and soft labels. More discussions can be found in Sect. 15. Similar findings were shown in recent research [2]. Besides, the soft label of a target sample is derived by the prediction of the source model, which is a class probability value. By leveraging the source data and the target data with hard and soft labels, the process of training the target model is as follows: (i) Train the source model w_s^k for device $k \in \mathcal{K}$ with $\mathcal{D}_s^k$; (ii) Use the learned source model to generate the soft label y_t^* for each sample $x_t \in \mathcal{X}_t$ in the target domain using softmax function σ. The soft label is defined by

$$y_t^* = \sigma(W_s(x_t)/T), \tag{1}$$

where W_s is the global source model by element-wise averaging local source model w_s^k for all device $k \in \mathcal{K}$ [19], and T is the temperature parameter to control the smoothness of the soft label. (iii) Train the target model w_t^k at device k using the hard and soft labels for each target data by

$$\arg\min \frac{1}{N_t^k} \sum_{i=1}^{N_t^k} [\lambda^k \ell_t(y_t^i, w_t^k(x_t^i)) + (1 - \lambda^k)\ell_t(y_t^{*i}, w_t^k(x_t^i))], \tag{2}$$

where N_t^k denotes the number of target domain samples at device k; ℓ_t is the loss function; w_t is the local target model; λ^k is the imitation parameter for device k to balance the importance between the hard label y_t and the soft label y_t^*.

Parallel Training Between Source and Target. Efficiency has been a well-known concern in distributed machine learning. Since both source and target data are distributed across devices, instead of waiting for a well-trained source model, we propose a parallel training paradigm to accelerate FSSDA. As shown in Algorithm 1 (line 6–10), each device trains the source and target model simultaneously. Although the source model does not perform well in the initial stage, it still promotes domain alignment and thus accelerates the generation of the target model. Thus the main purpose of parallel computing is to train the source and target models simultaneously, speeding up the overall training process. Our parallel design is empirically evaluated in the experiment section below. It should be mentioned that parallel training does not incur additional communication costs since target model updates can be appended to source updates.

Adaptive Imitation Parameters. Although the semi-supervised knowledge transfer module integrates SSDA and FL in Sect. 15, FSSDA suffers negative transfer from the noisy hard and soft labels. Specifically, due to limited labeling in the target domain (e.g., three labeled samples per class in the experiments), most hard labels are fake ones with limited ground truth knowledge, which restricts the domain alignment performance. Besides, the soft labels during parallel training upon the above module can be noisy during initial training. Due to the domain gap between source and target, even the well-trained source model may generate improper soft labels, and the entangled knowledge learned from the source may lead to serious negative transfer [22]. These problems become more challenging in federated settings, where target devices do not have access to any source domain data. To properly balance the importance between hard labels and soft labels, we develop an adaptive approach for selecting the imitation parameter λ in (2). Specifically, the imitation parameter controls how much knowledge can be transferred from the source domain, whose importance has been shown in prior KD research [9,17]. However, prior research determines the imitation parameter using either a brute-force search or domain knowledge, which cannot flexibly handle different domain discrepancies and noisy labels in FSSDA. Especially under heterogeneous federated settings, end devices have statistically heterogeneous data (non-iid) for both source and target domains.

To effectively select imitation parameters to adaptively use the noisy soft and hard labels, we consider problem (1) as a multi-task learning problem, where the soft loss and hard loss are the two task objectives. Since, in each federated training iteration, each device holds its own target domain data and the updated global source model, imitation parameters can be determined independently on the device side, which also addresses the data heterogeneity concern in federated settings. Specifically, we leverage the stochastic multi-subgradient descent algorithm (SMSGDA) [21], a well-known multi-task learning approach, to adaptively select the imitation parameter at each federated iteration for each individual device. The objective function can be given by

$$\min_{\lambda \in [0,1]} \|\lambda \nabla_w \ell_t(y_t, w_t(x_t)) + (1 - \lambda) \nabla_w \ell_t(y_t^*, w_t(x_t))\|^2, \tag{3}$$

where ℓ_t is the loss function, y_t is the hard label, and y_t^* is the soft label generated by the source model W_s for the target domain dataset. w_t is the local target model. The analytical solution to the above problem can be given by

$$\lambda = \nabla_w \ell_t(y_t^*, w_t(x_t)) \times \frac{(\nabla_w \ell_t(y_t^*, w_t(x_t)) - \nabla_w \ell_t(y_t, w_t(x_t)))^T}{\|\nabla_w \ell_t(y_t, w_t(x_t)) - \nabla_w \ell_t(y_t^*, w_t(x_t))\|^2}, \tag{4}$$

where λ is clipped between $[0,1]$. Therefore, each device can efficiently derive its local imitation parameter with the above closed-form solution.

3.3 FSSDA over Multi-source Domains

This part introduces the extension of the proposed FSSDA to multi-source scenarios. When the distributed source data includes multiple source domains, then it is essential to extract the inter-domain knowledge to align the domain-specific representations better. Define the total number of source domains by S. Thus, the overall learning objective at device $k \in \mathcal{K}$ for S source domains can be extended from (2) to

$$\arg \min \frac{1}{N_t^k} \sum_{i=1}^{N_t^k} [\lambda_1^k \ell_t(y_t^i, w_t^k(x_t^i)) + \sum_{j=1}^{S} \lambda_{j+1}^k \ell_t(y_t^{*ij}, w_t^k(x_t^i))], \tag{5}$$

$$s.t. \sum \lambda_i^k = 1,$$

where N_t^k is the total number of data samples in the target domain at device k, w_t is the local target model, y_t^{*ij} is the soft-label generated by the jth source model W_S^j for local data x_i, and λ^k is the imitation parameter for device $k \in \mathcal{K}$. In (5), imitation parameters are used to control more than two objective functions, i.e., in total $S + 1$ losses, to jointly optimize the target model. Thus, given the new condition for imitation parameters, problem (5) cannot be solved by the closed-form solution in (4). We propose to use the Frank-Wolfe-based optimizer to solve the constrained optimization, which can scale to high-dimensional problems with low computational overhead [10,28].

4 Experiments

4.1 Experimental Setup

We evaluate our models on the office dataset [26], which is widely used in domain adaptation. The office dataset includes 3 subsets: Webcam (795 samples) contains images captured by the web camera, Amazon (2, 817 samples) contains

Table 1. Performance comparison between FSSDA and baseline approaches. Six cases are considered between Amazon (A), Webcam (W), and DSLR (D).

	A → W	A → D	W → A	W → D	D → A	D → W
SSDAOnly (iid)	66.83%	66.10%	56.98%	75.67%	49.67%	73.37%
FLOnly (iid)	64.08%	70.10%	41.07%	70.10%	41.07%	64.08%
FSSDA (iid)	**83.01%**	**84.94%**	**66.23%**	**98.45%**	**71.39%**	**97.63%**
SSDAOnly (non-iid)	64.81%	59.39%	52.51%	69.80%	46.83%	69.33%
FLOnly (non-iid)	52.47%	63.65%	38.71%	63.65%	38.71%	52.47%
FSSDA (non-iid)	**82.15%**	**82.15%**	**66.09%**	**97.20%**	**69.67%**	**95.48%**

images downloaded from amazon.com, and DSLR (498 samples) contains images captured by a digital SLR camera, sharing 31 classes. In the following, we use W, A, and D to represent the above three subsets, respectively.

We consider both iid and non-iid data distributions in federated settings. We use the distribution-based label imbalance [16] to generate non-iid data distributions, where each end device is allocated a proportion of the samples whose labels follow Dirichlet distribution. Specifically, we sample $p_l \sim \text{DirN}(\beta)$ and allocate a $p_{l,k}$ proportion of the instances of class l to each device k. In our setting, we set the β value as 0.1. Besides, we consider practical SSDA settings, where limited labeled samples are given in the target domain. In iid and non-iid settings, only 93 labeled examples (3 per class) are distributed across all the end devices. We use ResNet-101 [13] for the baseline methods and the proposed method. All models are pre-trained on ImageNet [7]. The model parameters are optimized using stochastic gradient descent with a learning rate of 0.001.

For baseline approaches, existing SSDA requires access to data from different domains, which is ineffective in federated settings. Besides, none of the existing FL targets SSDA. Hence, to evaluate the proposed FSSDA, we consider two baseline approaches. (i) *SSDAOnly:* Without using FL, device local knowledge cannot be transferred due to privacy concerns. Each device performs SSDA to generate a local target model with its own data but does not participate in federated learning. (ii) *FLOnly:* Without effective SSDA in federated settings, end devices can only leverage labeled target data to learn the target model collaboratively. There is no knowledge transfer from the source domain.

4.2 Experimental Results

Effectiveness Evaluation. We consider six cases for domain adaptations between Amazon, Webcam, and DSLR under both iid and non-iid federated settings. As shown in Table 1, we observe that *FSSDA outperforms both SSDAOnly and FLOnly in all the cases.* We get the most promising result in the case of Webcam to DSLR both in iid and non-iid settings. The SSDAOnly and FLOnly get around 70%, whereas our proposed FSSDA methods achieve more than 97% accuracy. FLOnly cannot leverage the unlabeled samples, resulting in the

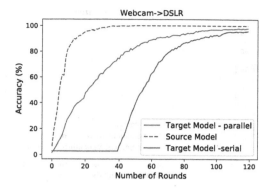

Fig. 2. Impact of the parallel training module of FSSDA (W → D).

worst performance in most cases. Although SSDAOnly leverages unlabeled target domains via knowledge transfer, SSDAOnly cannot utilize the shared knowledge from other end-devices, which makes the learning ineffective. Especially in the non-iid cases, the number of data varies for each end device; the performance degradation of one of the local models affects the aggregated global model. Both in iid and non-iid settings, DSLR as a target is able to achieve good performance of over 82% accuracy even when the domain gap is large (A → D). Moreover, due to the large domain gap between Webcam/DSLR and Amazon as well as the limited samples in Webcam/DSLR compared to Amazon, it is challenging to transfer knowledge to Amazon (W → A and D → A). However, we still achieve better results compared to baselines. SSDAOnly and FLOnly can only get around 50% accuracy, while the FSSDA can achieve accuracy close to 70%, demonstrating the effectiveness of FSSDA in challenging SSDA scenarios.

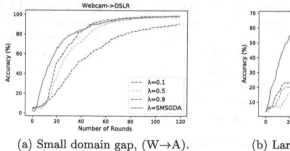
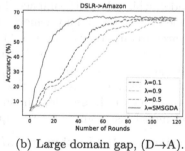

(a) Small domain gap, (W→A).   (b) Large domain gap, (D→A).

Fig. 3. Impacts of imitation parameters for FSSDA with different domain gaps.

Efficiency Evaluation. We compare the parallel training discussed in Sect. 3 with the serial training between the source (Webcam) and target (DSLR) models, as shown in Fig. 2. In serial training, the target model starts SSDA under

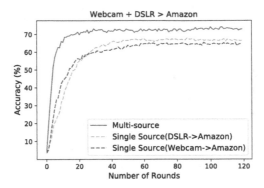

Fig. 4. FSSDA for multi-source SSDA (Target: Amazon).

federated settings until the source model is converged at around the 40th round. We observe that parallel training continuously outperforms serial training, which confirms the source model's positive impact on SSDA. Compared with serial training, the convergence rate of parallel training is significantly improved by around 33%.

Impact of Imitation Parameters. We illustrate the impact of the imitation parameter on FSSDA by using static and adaptive (SMSGDA) values. In our design, a large λ indicates learning more knowledge from the target domain (hard label) and less from the source domain (soft label), and vice versa. We conduct experiments under two different domain shift scenarios in Fig. 3: the small domain gap from Webcam to Amazon and the large domain gap from DSLR to Amazon. We observe that when the domain gap is large (Fig. 3(b)), at the initial stage, a lower value of the imitation parameter ($\lambda = 0.1$) will speed up the performance of the target model, but at the end, performance degrades, which shows the impact of negative transfer. Besides, a larger imitation parameter ($\lambda = 0.9$) finally achieves good accuracy but does not converge quickly compared to our adaptive design. From Fig. 3(a), when the domain gap is small, the negative transfer will not be significant ($\lambda = 0.1$), and thus we can able to rely more on the source domain. However, the proposed adaptive imitation parameter scheme is irrespective of the domain difference, which performs well for both small and large domain shifts. Overall, our adaptive design trains the target model faster and converges quickly.

Effectiveness Evaluation for Multi-source SSDA. We evaluate the performance of FSSDA under a multi-source scenario. We focus on FSSDA with Amazon as the target since Amazon has a large domain gap compared to the other two domains (DSLR and Webcam), which is the most challenging FSSDA setting under the office dataset. As shown in Fig. 4, multi-source FSSDA outperforms both single-source results from Webcam and DSLR, which demonstrates the effectiveness of our FSSDA in multi-source scenarios. Meanwhile, multi-source FSSDA further speeds up the overall federated training process to converge faster.

5 Conclusion

This paper proposed FSSDA to achieve semi-supervised domain adaptation (SSDA) over multiple distributed and confidential datasets. FSSDA integrates SSDA with federated learning based on adaptive and controllable knowledge transfer techniques, which include three key modules: semi-supervised knowledge transfer, parallel training, and adaptive imitation parameter selection. FSSDA can be used in single- or multiple-source SSDA problems. We empirically explored SSDA performance under iid and non-iid federated settings to validate the effectiveness and efficiency of our design.

Acknowledgement. The authors thank all anonymous reviewers for their insightful feedback. This work was supported by the National Science Foundation under Grants CCF-2106754, CCF-2221741, CCF-2153381, and CCF-2151238. The work of Zhen Liu was supported in part by Federal Highway Administration grant FHWA693JJ320-C000022, and the work of Xianhao Chen was supported in part by the HKU IDS Research Seed Fund under grant IDS-RSF2023-0012.

References

1. Albaseer, A., Ciftler, B.S., Abdallah, M., Al-Fuqaha, A.: Exploiting unlabeled data in smart cities using federated learning. arXiv preprint arXiv:2001.04030 (2020)
2. Ao, S., Li, X., Ling, C.: Fast generalized distillation for semi-supervised domain adaptation. In: Proceedings of the AAAI Conference on Artificial Intelligence, vol. 31 (2017)
3. Bagherinezhad, H., Horton, M., Rastegari, M., Farhadi, A.: Label refinery: Improving imagenet classification through label progression. arXiv preprint arXiv:1805.02641 (2018)
4. Chen, D., Mei, J.P., Wang, C., Feng, Y., Chen, C.: Online knowledge distillation with diverse peers. In: Proceedings of the AAAI Conference on Artificial Intelligence, vol. 34, pp. 3430–3437 (2020)
5. Cho, J.H., Hariharan, B.: On the efficacy of knowledge distillation. In: Proceedings of the IEEE/CVF International Conference on Computer Vision, pp. 4794–4802 (2019)
6. Daumé III, H., Kumar, A., Saha, A.: Frustratingly easy semi-supervised domain adaptation. In: Proceedings of the 2010 Workshop on Domain Adaptation for Natural Language Processing, pp. 53–59 (2010)
7. Deng, J., Dong, W., Socher, R., Li, L.J., Li, K., Fei-Fei, L.: Imagenet: a large-scale hierarchical image database. In: 2009 IEEE Conference on Computer Vision and Pattern Recognition, pp. 248–255. IEEE (2009)
8. Donahue, J., Hoffman, J., Rodner, E., Saenko, K., Darrell, T.: Semi-supervised domain adaptation with instance constraints. In: Proceedings of the IEEE Conference on Computer Vision and Pattern Recognition, pp. 668–675 (2013)
9. Duan, L., Xu, D., Tsang, I.: Learning with augmented features for heterogeneous domain adaptation. arXiv preprint arXiv:1206.4660 (2012)

10. Frank, M., Wolfe, P.: An algorithm for quadratic programming. Naval Res. Logist. Q. **3**(1–2), 95–110 (1956)
11. Furlanello, T., Lipton, Z., Tschannen, M., Itti, L., Anandkumar, A.: Born again neural networks. In: International Conference on Machine Learning, pp. 1607–1616. PMLR (2018)
12. Ganin, Y., et al.: Domain-adversarial training of neural networks. J. Mach. Learn. Res. **17**(1), 2030–2096 (2016)
13. He, K., Zhang, X., Ren, S., Sun, J.: Deep residual learning for image recognition. In: Proceedings of the IEEE Conference on Computer Vision and Pattern Recognition, pp. 770–778 (2016)
14. Hinton, G., Vinyals, O., Dean, J.: Distilling the knowledge in a neural network. arXiv preprint arXiv:1503.02531 (2015)
15. Jeong, W., Yoon, J., Yang, E., Hwang, S.J.: Federated semi-supervised learning with inter-client consistency & disjoint learning. arXiv preprint arXiv:2006.12097 (2020)
16. Li, Q., Diao, Y., Chen, Q., He, B.: Federated learning on non-IID data silos: an experimental study. arXiv preprint arXiv:2102.02079 (2021)
17. Lopez-Paz, D., Bottou, L., Schölkopf, B., Vapnik, V.: Unifying distillation and privileged information. arXiv preprint arXiv:1511.03643 (2015)
18. Luo, P., Zhu, Z., Liu, Z., Wang, X., Tang, X.: Face model compression by distilling knowledge from neurons. In: Thirtieth AAAI Conference on Artificial Intelligence (2016)
19. McMahan, B., Moore, E., Ramage, D., Hampson, S., Arcas, B.A.: Communication-efficient learning of deep networks from decentralized data. In: Artificial Intelligence and Statistics, pp. 1273–1282 (2017)
20. Meng, Z., Li, J., Gong, Y., Juang, B.H.: Adversarial teacher-student learning for unsupervised domain adaptation. In: 2018 IEEE International Conference on Acoustics, Speech and Signal Processing (ICASSP), pp. 5949–5953. IEEE (2018)
21. Milojkovic, N., Antognini, D., Bergamin, G., Faltings, B., Musat, C.: Multi-gradient descent for multi-objective recommender systems. arXiv preprint arXiv:2001.00846 (2019)
22. Pan, S.J., Yang, Q.: A survey on transfer learning. IEEE Trans. Knowl. Data Eng. **22**(10), 1345–1359 (2009)
23. Peng, X., Huang, Z., Zhu, Y., Saenko, K.: Federated adversarial domain adaptation. arXiv preprint arXiv:1911.02054 (2019)
24. Peterson, D., Kanani, P., Marathe, V.J.: Private federated learning with domain adaptation. arXiv preprint arXiv:1912.06733 (2019)
25. Polino, A., Pascanu, R., Alistarh, D.: Model compression via distillation and quantization. arXiv preprint arXiv:1802.05668 (2018)
26. Saenko, K., Kulis, B., Fritz, M., Darrell, T.: Adapting visual category models to new domains. In: Daniilidis, K., Maragos, P., Paragios, N. (eds.) ECCV 2010. LNCS, vol. 6314, pp. 213–226. Springer, Heidelberg (2010). https://doi.org/10.1007/978-3-642-15561-1_16
27. Saito, K., Kim, D., Sclaroff, S., Darrell, T., Saenko, K.: Semi-supervised domain adaptation via minimax entropy. In: Proceedings of the IEEE/CVF International Conference on Computer Vision, pp. 8050–8058 (2019)
28. Sener, O., Koltun, V.: Multi-task learning as multi-objective optimization. In: Advances in Neural Information Processing Systems, vol. 31 (2018)
29. SOBERS, R.: 98 must-know data breach statistics for 2021 (2021). https://www.varonis.com/blog/data-breach-statistics/

30. Urban, G., et al.: Do deep convolutional nets really need to be deep (or even convolutional)? (2016)
31. Wang, M., Deng, W.: Deep visual domain adaptation: a survey. Neurocomputing **312**, 135–153 (2018)
32. Yang, Q., Liu, Y., Chen, T., Tong, Y.: Federated machine learning: concept and applications. ACM Trans. Intell. Syst. Technol. (TIST) **10**(2), 1–19 (2019)
33. Yao, T., Pan, Y., Ngo, C.W., Li, H., Mei, T.: Semi-supervised domain adaptation with subspace learning for visual recognition. In: Proceedings of the IEEE Conference on Computer Vision and Pattern Recognition, pp. 2142–2150 (2015)
34. Yim, J., Joo, D., Bae, J., Kim, J.: A gift from knowledge distillation: fast optimization, network minimization and transfer learning. In: Proceedings of the IEEE Conference on Computer Vision and Pattern Recognition, pp. 4133–4141 (2017)
35. Zhang, L., Yuan, X.: Fedzkt: zero-shot knowledge transfer towards heterogeneous on-device models in federated learning. arXiv preprint arXiv:2109.03775 (2021)
36. Zhou, K., Yang, Y., Qiao, Y., Xiang, T.: Domain adaptive ensemble learning. IEEE Trans. Image Process. **30**, 8008–8018 (2021)
37. Zhu, X., Goldberg, A.B.: Introduction to semi-supervised learning. Synth. Lect. Artif. Intell. Mach. Learn. **3**(1), 1–130 (2009)
38. Zhuang, W., Gan, X., Wen, Y., Zhang, X., Zhang, S., Yi, S.: Towards unsupervised domain adaptation for deep face recognition under privacy constraints via federated learning. arXiv preprint arXiv:2105.07606 (2021)

A Lightweight Reputation System for UAV Networks

Simeon Ogunbunmi[1], Mohsen Hatmai[1], Ronghua Xu[2], Yu Chen[1(✉)],
Erik Blasch[3], Erika Ardiles-Cruz[3], Alexander Aved[3], and Genshe Chen[4]

[1] Binghamton University, Binghamton, NY 13902, USA
ychen@binghamton.edu
[2] Michigan Technological University, Houghton, MI 49931, USA
[3] The U.S. Air Force Research Laboratory, Rome, NY 13441, USA
[4] Intelligent Fusion Tech, Inc., Germantown, MD 20876, USA

Abstract. Unmanned Aerial Vehicles (UAVs) have become indispensable components in the modern Internet of Things (IoT) ecosystem and are increasingly popular for various applications, including delivery, transporting, inspection, and mapping. However, the reliability, security, and privacy of UAV devices are among the public's top concerns as they operate close to each other and other objects. This paper proposes a **LI**ghtweight **B**lockchain-based **RE**putation (LIBRE) system to improve the reliability and performance of a UAV network by monitoring, tracking, and selecting the most appropriate individuals to carry out tasks. In the LIBRE system, a reputation score is assigned to each newly registered UAV device with limited network access. Exclusive access is, therefore, given once the reputation is ascertained based on the behavior and the feedback given by peer nodes that have interacted with it. An algorithm was proposed to calculate the reputation score updated in the Blockchain to provide fairness, immutability, and auditability. A proof-of-concept prototype of LIBRE system architecture was implemented on a private Ethereum Blockchain, and the extensive experimental study has validated the effectiveness of the LIBRE scheme.

Keywords: Unmanned Aerial Vehicles (UAVs) · Reputation System · Reliability · Lightweight · Ethereum Blockchain

1 Introduction

From 2018 to 2023, the market of Unmanned Aerial Vehicles (UAVs), which are also referred to as Drones, has grown from 69 billion dollars to 141 billion dollars

This work was partially supported by the U.S. National Science Foundation (NSF) under Grant No. 2141468, and the U.S. Air Force Research Laboratory (AFRL) Summer Faculty Fellowship Program (SFFP) via contracts FA8750-15-3-6003, FA9550-15-001, and FA9550-20-F-0005. The views and conclusions contained herein are those of the authors and should not be interpreted as necessarily representing the official policies or endorsements, either expressed or implied, of the U. S. Air Force.

ⓒ ICST Institute for Computer Sciences, Social Informatics and Telecommunications Engineering 2024
Published by Springer Nature Switzerland AG 2024. All Rights Reserved
Y. Chen et al. (Eds.): SmartSP 2023, LNICST 552, pp. 114–129, 2024.
https://doi.org/10.1007/978-3-031-51630-6_8

[13]. Because of many attractive features, drones have been widely adopted for both civilian and military applications including delivery, transporting, inspection, surveillance, and mapping [5,7,17]. In particular, data received through the UAV devices is essential and crucial for carrying out emergency operations [2]. Different from other robotic vehicles, UAVs have demonstrated higher mobility and adaptivity, which are required for tasks conducted in remote worksites, inconvenient or hazardous locations, or areas that lack communication infrastructure [29], leaving alone the capability of collecting high-quality images for complex tasks. Essentially, each UAV unit is a complex Internet of Things (IoT) system, which consists of sensors, antennae, embedded software, and a two-way communication module. The whole system functions seamlessly to ensure the Quality of Service (QoS) and the Quality of Experience (QoE) for applications like remote control and monitoring [9,27].

The proliferation of UAV applications also made UAV networks attractive targets. The past decades have witnessed variant attacks against UAV systems on confidentiality, reputation, privacy, security, and reliability [1]. Compared to regular computers, UAV systems are more vulnerable to physical and cyber threats due to their constrained computing resources and limited power supply [13]. To manage behavioral evidence and enforce authorization, an effective access control layer is required in addition to authentication [4]. During encrypted data exchange between UAV systems and unauthorized entities, sensitive and private information, such as position, payload, and flight time, is made public [11], making them highly vulnerable to attacks.

The reliability and credibility of UAV networks are of paramount importance. A reputation system aggregates the interactions among nodes and enables the establishment of profiles that reflect system-level and individual-level performance [10]. Reputation scores serve as indicators of security, privacy, and decision-making confidence, providing valuable insights into the level of trustworthiness. Since its introduction as a decentralized and transparent ledger technology with characteristics such as audibility, immutability, traceability, and transparency, Blockchain has garnered recognition as a promising solution to enhance privacy and security in data transmission [6,27]. By incorporating peer-to-peer and cryptography consensus algorithms, Blockchain has achieved transparency characteristics between different non-trusted entities [22].

This paper introduces a **LI**ghtweight **B**lockchain-based **RE**putation (LIBRE) system to enhance the reliability and performance of UAV networks. LIBRE achieves this by monitoring, tracking, and selecting the most suitable individuals for task execution. This protocol effectively eliminates malicious nodes from the UAV network, allowing the network to reach a consensus when assessing the overall system reliability [16]. The major contributions are listed below:

- A light-weight reputation system architecture is introduced with the details of its key components and functionalities;
- A reputation contract to determine malicious or harmful devices that cause attacks on the system was proposed; and
- A proof-of-concept prototype of LIBRE architecture is implemented and tested, which validated the proposed architecture.

The rest of the paper is structured as follows: Sect. 2 provides a concise overview of the background knowledge and related works on UAV networks and reputation systems. The proposed LIBRE design rationale and architecture are presented in Sect. 3, and the experimental results are reported in Sect. 4. Finally, Sect. 5 concludes the paper with a brief discussion of the ongoing efforts.

2   Background and Related Works

This section provides a brief overview of the background knowledge and related works on Blockchain networks for UAVs, and the implementation of reputation systems using different frameworks, models, and consensus protocols.

2.1   Blockchain in UAV Networks

Blockchain is based on a distributed database with a scalable list of data entries. The information block includes the timestamp, encrypted hash value, and data transaction from the preceding linked block [21]. In a blockchain network, when nodes exchange information or assets, they initiate a transaction. The source node creates a transaction file, which is broadcast to the network or specific nodes for validation. Validated transactions are grouped into blocks and added to the blockchain based on the consensus mechanism employed [20]. Numerous researchers and organizations have contributed to the creation and improvement of blockchain technology since it was first introduced with Bitcoin. Building on the outlining of the original concept, there are variant blockchain systems and versions that have been developed by diverse researchers today. Blockchain technology has tremendous potential in various fields where trust is essential between mutually dependent parties. Its applications include not just electronic cash exchange systems like Bitcoin and Litecoin [3], but also rendering and enabling secure communication amongst robotic swarm systems or even data marketplaces [26]. There is sufficient literature that covers the taxonomy on the use of blockchain for authentication in IoT networks [19] and the challenges of adopting blockchain in IoT devices alongside some of their solutions to these challenges [18].

   Specifically, the integration of Blockchain technology has proven effective in mitigating various attacks in UAV systems, including Sybil attacks, Man-in-the-Middle attacks, jamming, Distributed Denial of Service (DDoS) attacks, and more [25]. Integrating blockchain enables establishing trust and ensuring data immutability and transparency within UAV systems. Several studies have highlighted the effectiveness of Blockchain in enhancing security and integrity. The Autonomous Intelligent Robot Agent (AIRA) protocol is introduced to address the limitations of a centralized system [15]. This protocol utilizes blockchain, specifically the Ethereum platform, to manage economic interactions in a multi-agent system. AIRA protocol combines smart contracts, Robot Operating System, InterPlanetary File System for data storage and Docker for virtualization. Transactions in the system involve both Ethereum tokens and custom tokens

[15]. The AIRA protocol was implemented in the Drone Employee project, where UAVs were utilized for navigation, regulatory compliance, and economic activities. The process involved service requests, smart contract creation, service acceptance by UAV agents, and approval of air corridors by agent dispatchers [15].

To address the constraints of IoT devices and support UAV-based applications to securely and autonomously receive sensor data, a decentralized platform within the air-to-ground heterogeneous network is suggested [12], which enables information storage and trading. A novel blockchain architecture is introduced that effectively addresses computation and storage overhead while maintaining privacy, security, and lightweight characteristics [12].

The significance of blockchain in the context of UAV-assisted IoT is highlighted and a data collection system is proposed in [28], which emphasizes security and energy efficiency. Blockchain is introduced as a fundamental component that enables UAVs to serve as edge data collection nodes. By leveraging blockchain technology, the UAVs facilitate long-term network access for IoT devices through regular cruises and recharging [28]. This integration of blockchain with UAV-assisted IoT showcases the importance of blockchain in creating a comprehensive framework that incorporates UAV edge computing, UAV charging, and secure data handling.

Focusing on the common security and privacy concerns in IoT, a framework is proposed that combines blockchain with IoT to address these issues effectively [30]. By integrating blockchain technology, the framework offers robust security and privacy measures, ensuring the integrity of IoT data and supporting various functionalities such as authentication and decentralized payment. Potential solutions are also presented based on blockchain and Ethereum to tackle security challenges in IoT devices [30], including data sharing, data integrity, authentication, access control, and privacy. The use of blockchain serves as a promising solution to enhance the overall security of IoT systems.

Blockchain technology plays a crucial role in addressing the issues associated with centralized solutions by introducing the UGG/IPP and LPP algorithms for dynamic encryption [21]. A decentralized architecture is proposed that leveraged hash functions to enhance storage and processing efficiency utilizing blockchain. The significance of blockchain in this context lies in its capability to provide a secure and tamper-resistant framework for storing and managing identities. The proposed architecture, with its periodic updates and calculation of real identities, showcased improved system performance, reduced processing time, and enhanced privacy protection [21].

2.2 Reputation Systems

Trust and reputation systems are critical in a variety of fields, including online platforms, social networks, and distributed systems. They provide for the evaluation of the trustworthiness and trust of entities such as users, service providers, or peers based on their previous behavior, interactions, and feedback. Researchers have extensively researched the design, analysis, and assessment of trust and reputation systems in journal publications to improve security, minimize assaults, and improve decision-making processes.

A trust and reputation model is crucial for protecting large distributed sensor networks in IoT/CPS from malicious node attacks. Such a model fosters collaboration among distributed entities, aids in detecting untrustworthy entities, and assists in decision-making processes. After thoroughly exploring trust establishment processes and comparing various methods, a trust and reputation model called TRM-IoT is designed to promote cooperation among IoT/CPS network things based on their behaviors [8]. The model's accuracy, robustness, and efficiency are validated through extensive simulations, demonstrating its effectiveness in ensuring reliable and lightweight trust management in IoT/CPS networks [8].

Researchers also proposed solutions for trust and reputation systems based on Fog computing [23]. It utilizes Fog nodes to evaluate trust levels among IoT devices, allowing interactions only with devices that meet a predefined trust threshold. This evaluation process helps to prevent malicious devices from impacting the system and compromising the quality of service while also safeguarding against various attacks such as Bad Mouthing, On-Off, and Self Promoting attacks [23]. The paper includes simulation results demonstrating the system's behavior under these attacks. Additionally, the proposed solution is well-suited for large-scale IoT systems. A comparison with related works reveals that the proposed model outperforms previous approaches in terms of its suitability for IoT systems and security.

An event-based reputation model is introduced aimed at filtering false event messages in a multi-UAV network [14]. The proposed solution recognized two distinct roles for each event and implemented a dynamic mechanism for role development, reputation, and evaluation. The mechanism helped to determine the trustworthiness of incoming messages and prevents the spread of false event messages among UAVs in the network [14]. By employing this approach, the system can effectively mitigate the impact of false information and maintain the reliability of event communication in the multi-UAV environment.

An enhanced condensed hierarchical clustering method was proposed that utilizes user preference similarity to enhance the accuracy of recommendation trust [24]. This approach employed a cloud model-based technique to measure similarities between users and then applied a hierarchical clustering method to group users into different domains based on their similarities. This process obtained the final recommendation trust, which includes both intra-domain and extra-domain recommendation trust [24]. The overall trust in cloud services is evaluated by considering both direct trust and recommended trust. Through simulation experiments, the paper validated the accuracy and superiority of the clustering algorithm. The experimental results demonstrated that the cloud service selection method enhances transaction success rates and allows users to choose more satisfactory cloud services.

3 LIBRE: Rationale and Architecture

Aiming at assurance and reliability of UAV systems, LIBRE leverages reputation system and Blockchain technology to enhance QoS and security requirements in drone-based applications, like package delivery, smart surveillance, environment monitoring, etc. Figure 1 demonstrates the LIBRE system architecture that consists of four sub-systems: i) UAV network; ii) identity authentication; iii) reputation system; and iv) Blockchain fabric.

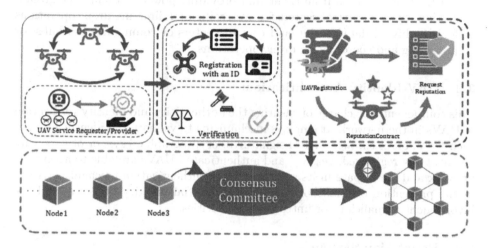

Fig. 1. LIBRE System Architecture Overview.

3.1 The UAV Network

The UAV network serves as a physical infrastructure for LIBRE system, which allows UAVs to connect, share data, and carry out specific functions to provide intelligent air mobility applications. Key components and functions are described as follows.

– *Unmanned Aerial Vehicles*: Unmanned Aerial Vehicles (UAVs) also known as drones are responsible for sensing the environment, collecting data, and communicating with ground stations. Unmanned Aerial Vehicles (UAVs) have the potential to radically improve speed, safety, and integration while completely transforming communication and transportation networks. Drones are UAVs that execute activities and services on the network. However, current events have shown how vulnerable UAVs are to attacks made feasible by faulty or malicious equipment operating within communication networks. To protect UAVs in the airspace and lessen the risk of cyber attacks, this emphasizes the urgent need for cybersecurity measures. Introducing a secure and reliable networking architecture for UAV data, blockchain is a concept that addresses this.

– *UAV Service Providers*: UAV Service Providers are nodes that provide services needed in the system and have the node registered on the network by interacting with the registration contract. They are responsible for the maintenance and operation of the UAVs, as well as the provision of flight services to users. They are used in making UAVs to perform tasks such as aerial photography, surveillance, and delivery.
– *Ground Stations*: The ground stations are the stationary points on the ground that control the UAVs. They are in charge of delivering commands to the UAVs, receiving data from them, and providing power to them. The ground stations serve as control centers for overseeing and directing drone operations for simplifying data sharing and control centers or communication hubs for managing and coordinating drone operations.

3.2   UAV Identity Authentication:

This component is in charge of ensuring the authenticity and identity verification of UAVs in the network. It employs mechanisms to validate the identification of each UAV, often through digital signatures, cryptographic keys, or other secure means. Only registered, verified, and authenticated UAVs are able to access the network and participate in its activities. This authentication mechanism assists in the prevention of unauthorized access, potential security breaches, and the involvement of malicious or untrustworthy entities.

3.3   Reputation System

The reputation system is critical to ensuring quick and secure service exchanges inside the UAV network. Because of their demonstrated track record of dependability and competency, providers with higher reputation scores are most likely to be preferred for services. By assessing and regulating the dependability and performance of UAVs, the reputation system is a fundamental component of the architecture that fosters reliability and accountability among UAV network members. Its purpose is to evaluate and maintain the reputation of the UAV service providers based on their behavior, dependability, and adherence to network rules. It tracks each UAV's activities and performance and assigns reputation ratings based on their actions and results. Provider reputation ratings may reflect factors such as performance, responsiveness, task completion, interactions, honesty, promptness in executing obligations, and adherence to safety regulations.

3.4   Blockchain

Blockchain is a decentralized network that provides a space where no one organization has total authority. This enhances the network's fault tolerance and resilience since different drones may communicate and cooperate without depending on a centralized authority. Additionally, blockchain provides trust and transparency through its transparent and immutable ledger. In a drone network, this ensures secure recording of flight data, including location, altitude, and mission parameters. Transparency builds trust among network participants and safeguards the integrity of the collected data.

3.5 Architecture and Algorithm

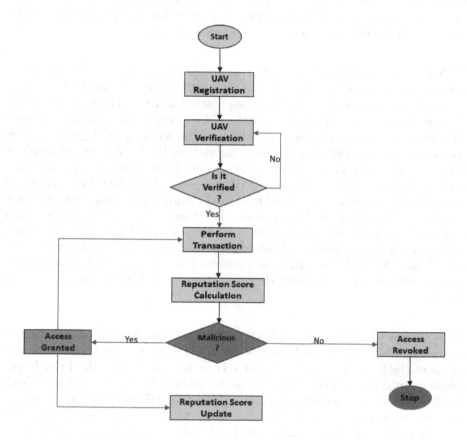

Fig. 2. The process of gaining access to a network.

To manage the registration, verification, reputation calculation, and updating of the reputation scores for service providers in UAV systems, the LIBRE system architecture was implemented. With Remix IDE using Solidity version 0.8.20, the solidity code was compiled using Remix VM (Shangbai), and the Reputation Calculation Equation (Eq. 1) was used in this design to calculate the reputation score of each device after each successful transaction made by each provider based on their weight factors, rating and the number of tasks completed. Separate smart contracts were developed to ensure separate reputation operations for the UAV providers. For service providers, the ReputationSystem contract maintains the reputation system and reputation management.

The ReputationContract contract provides a different approach to determining reputation, taking into account elements like ratings, tasks completed, and weight into consideration. Two interfaces were implemented by the smart contracts to outline the functions of each smart contract used. the IReputationSystem which was implemented by the Reputation Contracts, while the IUAVRegistration was implemented by the UAVRegistration contracts. The functions of the smart contracts used for registration, verification and reputation calculations, and updating in the design are discussed as follows:

UAVRegistration Contract: The reputation system interface (IReputationSystem) is used in the UAVRegistration contract to implement the UAV registration, and verification, and initiate the reputation score updating process. From the reputation system contract, an initial UAV's reputation score is generated. This smart contract allows the registration of the service provider with a given specific name using a string parameter representing the UAV to be registered. It is important to note that an empty name can not be accepted as a specific name needs to be given to each registered UAV. It also checks if the UAV is registered or verified by the name and the address used during registration. Once the registration and verification of the provider are completed, it sets the reputation score and the prior reputation to zero which is mapped using the name and address as the key. It returns a boolean output of the verification function to be true if the UAV device is verified and returns false if not verified. Transactions can not be done without a complete verification of the UAV device as shown in Fig. 2.

Reputation Contract: The reputation score of a device can be increased or decreased, which substantially affects the reputation, trustworthiness and the scores of the UAV devices either positively or negatively after the Prior Reputation calculation has been calculated as regarded in Eq. (1). Reviews/feedback are given after each successful transaction based on different metrics, e.g., delivery level, performance, delivery time, meeting the set rules and regulations, and reliability while we set our metrics to be timeliness and quality of delivery as we assume that the quality of the service deliver is assumed to be highly effective than the timeliness of the delivery, thereby, the weight factor of quality of delivery and timeliness of delivery is assumed to be 6 and 4 respectively. The feedback given is based on the rating and weight factors and it is calculated and updated using the reputation contract. The Reputation Score of each device is therefore calculated using the equation in Eq. (1) by finding the summation of all the prior reputation scores as regarded in Eq. (1) and then finding the division of this sum with the total amount of task it has completed.

$$PR = \frac{(WQ * RQ) + (WT * RT)}{10} \tag{1}$$

Algorithm 1. UAVRegistration

1: registered_names ← {}
2: verified_uavs ← {}
3: **procedure** REGISTERUAV($name$)
4:    **if** $name \neq$ None **and** $name \notin$ registered_names **and** $name \notin$ verified_uavs **then**
5:        registered_names[$name$] ← False
6:    **end if**
7: **end procedure**
8: **procedure** VERIFYUAV($name$)
9:    **if** $name \in$ registered_names **and** $name \notin$ verified_uavs **then**
10:        verified_uavs[$name$] ← True
11:        **delete** registered_names[$name$]
12:    **end if**
13: **end procedure**
14: **function** ISUAVREGISTERED($name$)
15:    **return** $name \in$ registered_names
16: **end function**
17: **function** ISUAVVERIFIED($name$)
18:    **return** $name \in$ verified_uavs
19: **end function**

$$RS = \sum_{1}^{n} \frac{PR}{TT} \qquad (2)$$

where:

WQ: the weight factor of Quality of Delivery

RQ: the reputation score of Quality of Delivery

WT: the weight factor of Timeliness

RT: the reputation score of Timeliness

TT: the Total Task Completed

PR: the Prior Reputation

RS: the Reputation Score

A node with a lower reputation score indicates that it has the potential or has already produced services that cause attacks and harm to the system, thereby the access of the node or such service provider is revoked to prevent malicious attacks on the system. A malicious provider can pass through the registration and verification process without being detected as malicious but can be detected once service is been rendered on the system. The calculated reputation score is then updated and saved on the blockchain to ensure the immutability, transparency, and decentralization of the system.

Algorithm 2. ReputationContract

1: **struct** UAV
2: **struct** Rating
3: **mapping(address = UAV)**
4: **mapping(address = Rating[])**
5: **public** reputationScores;
6: **mapping(address = uint256) public** priorReputations;
7: **constant**$MAX\_SCORE = 5$;
8: **constant**$MAX\_RATING = 5$;
9: **constant**$MAL\_THRESHOLD = 2$;
10: **constant**$WEIGHT\_QUALITY = 6$;
11: **constant**$WEIGHT\_TIMELINESS = 4$;
12: **function** SUBMITRATING(...)
13: **end function**
14: **function** CALCULATEREPUTATIONSCORE(...)
15: **end function**

---

4   Experimental Results

4.1   Experimental Setup

The algorithms were tested using multiple scenarios of UAV systems and various outputs based on the algorithm provided in Sect. 4. This is done with the Remix IDE environment, a web-based integrated development environment, that was employed for writing, testing, and deploying the smart contracts, which gives valuable logs for checking the status and results of each operation when debugged. The simulation was done on the Windows 10 operating system. Python 3 is the programming language employed in this implementation. A 500 GB SSD drive was used to meet storage requirements, ensuring quick access to data and low latency during operations and a 1 Gbps network link enabled flawless communication between nodes. Ethereum, a well-known and widely utilized blockchain platform, served as the foundation technology for building and testing our smart contracts. In order to simulate the real-time interaction inherent to blockchain-based systems, nodes interacted with one another through the use of smart contracts issued on the Ethereum blockchain. Some of our nodes served as Service Providers providing UAV services, while others were allocated specialized duties. Five nodes were selected for the simulation and the reputation score was computed by taking into account the nodes shown in Fig. 3. The reputation values for the 5 nodes are calculated using Equations (1) and (2) with a reputation score of 5 being considered as being reliable and reputation score below 3 is considered as malicious to the system and sending fake and wrong messages to the network.

4.2   Results

After implementing the framework, precise results of a node's reputation are produced, which may be used to assess whether or not the node is to be trusted.

According to Fig. 3, Node 1 and Node 2 have a balanced reputation in that they provide both real and fake services in a balanced manner. Also, Node 3's reputation is continually being stabilized as a result of its constant genuine service to the system with a constant rating being given to it. Because of its unreal services, Node 5 reputation is continually deteriorating and unreal. However, Node 4 has a fluctuating reputation score as a result of having to deliver both real and fake services to the system making the curves decrease as well as increase. The level at which the communications are regarded trustworthy is assumed to be 3.0, and relevant steps are performed. Devices that have a reputation score below this limit are deemed untrustworthy and malicious to the system, thereby they are discarded so as to prevent an attack on the system. Whether the node is accepted or rejected, the reputation of each node is updated as shown in the flow chart in Fig. 2. The closest the reputation score is to 5 which is assumed to be the optimum reputation score, the most trustworthy the device is.

The processing time for registration, verification, and rating process of each Unmanned Aerial Vehicle are plotted in Fig. 4. The graph shows that the processing time of UAVs during registration takes an average of 3 s to be completed for all the UAVs. This illustrates that differences in UAV models or other parameters have no significant effect on the registration process. A constant processing time suggests that the registration system is well-designed and efficient for the model. With an average processing time of 3 s, the registration procedure for all UAVs is fast. An efficient registration system is essential for UAV operations since it allows for quick deployment and eliminates downtime. It would also be beneficial to compare the processing times of the verification and rating proce-

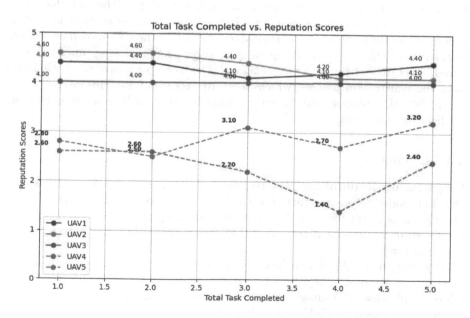

Fig. 3. Total task completed versus Reputation score

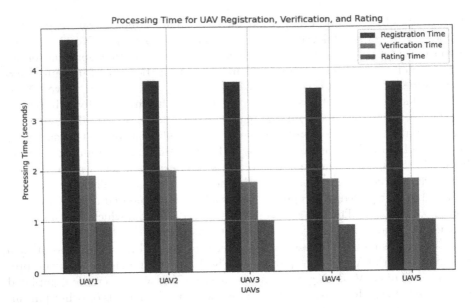

Fig. 4. Processing time (second) versus UAVs

dures because they are both consistent and efficient. This can help enhance and identify possible bottlenecks in UAV operations.

The verification technique takes an average of 2 s for each of the 5 UAVs, whereas the rating procedure takes 1 s according to Fig. 4. The fact that these processing times are minimal and quick indicates that the verification and rating processes are both efficient. The timely completion of these tasks can contribute to the overall efficiency of UAV operations. When the processing times are compared, we can see that the verification and rating operations are both faster than the registration process, which takes an average of 3 s. This suggests that the registration process may be more complicated or include more steps than verification and rating.

4.3 Discussions

Registration Reputation. The requester node gives feedback on the service provider nodes it has interacted with. This approach will fail if the service provider is not registered in the smart contract. Furthermore, the smart contract's state is restored if the submitted reputation values are outside of the expected range.

However, if a UAV device has already been registered and verified on the system, any attempt to register the account again either by its name or by the address will cause the transaction to be reversed to its initial state while printing out an error message "UAV is already registered".

When a UAV device with no prior registration history aims at performing some transactions on the network, a reversion to its initial state occurs indicating that the UAV has not been registered.

ReputationContract Reputation. The aggregation of the past feedback and present feedback after being calculated using the reputation calculation was computed and updated on the Reputation Contract. This node acted in good behavior by delivering honest and consistent evaluations in the majority of its reputation score submissions.

Figure 3 shows that after the rating was submitted for each UAV the reputation score was calculated. Plotting the reputation score per total task completed, it shows the interaction between the service providers after each successful transaction which shows the behaviors of both the malicious devices and the non-malicious devices.

Challenges of Blockchain on the UAV Network. Integrating blockchain into UAV networks is still facing many challenges, even if a lightweight-designed blockchain like Microchain. In ongoing efforts, we continue tailoring the blockchain to fit in the resource-constrained UAVs from aspects below:
Energy Efficient: UAV networks are resource-constrained devices Blockchain computational processes are intensive, and a large amount of energy is being consumed during computations most especially for UAV devices operating in real-time scenarios leading to the draining of the battery energy as fast as possible.
Connectivity: Most UAV operations occur in remote areas or hostile environments. Thereby connection to the blockchain networks in these environments continuously is challenging which can enhance synchronization issues.
Storage: Resource constraint features of UAV devices make it an issue when considering the storage of the devices. Having to store blockchain ledger on these resource-constrained devices is a trade-off and seems impractical as a result of the limited storage capacity which might hinder the up-to-date update of the blockchain.
Scalability: As a result of the huge amount of data being sent and received in a real-time scenario of UAV systems, scalability is a great issue on blockchain networks because if the transactions load, and may find it hard to handle this amount of data or load.

5 Conclusions

In this paper, we proposed a **LI**ghtweight Blockchain-based **RE**putation (LIBRE) system to enhance the reliability and performance of UAV networks. LIBRE system enables a UAV network to register, verify assignments, and update the reputation score of a UAV service provider after assessing and observing its behavior and services provided based on the feedback received. Reputation scores are calculated using the blockchain-based algorithm that guarantees the reliability, immutability, and auditability of the score after being updated. Based on the output of the reputation score calculated using the Reputation Calculation Equation, there is either an increase or decrease in the global reputation score which is visible to the public when the calculation and updating has been done. Based on the reliability, security, and privacy of UAV networks have been great issues being researched. A blockchain which is a type of lightweight distributed ledger technology system that uses fewer nodes to ensure faster consensus and transaction processing was implemented in the system as an underlying protocol for the reputation system. This lightweight method benefits UAV systems since it concentrates on systems with fewer resources.

References

1. Abdelmaboud, A.: The internet of drones: requirements, taxonomy, recent advances, and challenges of research trends. Sensors **21**(17), 5718 (2021)
2. Alladi, T., Chamola, V., Sahu, N., Guizani, M.: Applications of blockchain in unmanned aerial vehicles: a review. Veh. Commun. **23**, 100249 (2020)
3. Bansal, G., Hasija, V., Chamola, V., Kumar, N., Guizani, M.: Smart stock exchange market: a secure predictive decentralized model. In: 2019 IEEE Global Communications Conference (GLOBECOM), pp. 1–6. IEEE (2019)
4. Battah, A.A., Iraqi, Y., Damiani, E.: A trust and reputation system for IoT service interactions. IEEE Trans. Netw. Serv. Manage. **19**(3), 2987–3005 (2022)
5. Bhoi, S.K., Jena, K.K., Jena, A., Panda, B.C., Singh, S., Behera, P.: A reputation deterministic framework for true event detection in unmanned aerial vehicle network (UAVN). In: 2019 International Conference on Information Technology (ICIT), pp. 257–262. IEEE (2019)
6. Bodkhe, U., et al.: Blockchain for industry 4.0: a comprehensive review. IEEE Access **8**, 79764–79800 (2020)
7. Caro, M.P., Ali, M.S., Vecchio, M., Giaffreda, R.: Blockchain-based traceability in agri-food supply chain management: a practical implementation. In: 2018 IoT Vertical and Topical Summit on Agriculture-Tuscany (IOT Tuscany), pp. 1–4. IEEE (2018)
8. Chen, D., Chang, G., Sun, D., Li, J., Jia, J., Wang, X.: TRM-IoT: a trust management model based on fuzzy reputation for internet of things. Comput. Sci. Inf. Syst. **8**(4), 1207–1228 (2011)
9. Chen, N., Chen, Yu.: Smart city surveillance at the network edge in the era of IoT: opportunities and challenges. In: Mahmood, Z. (ed.) Smart Cities. CCN, pp. 153–176. Springer, Cham (2018). https://doi.org/10.1007/978-3-319-76669-0_7
10. Debe, M., Salah, K., Rehman, M.H.U., Svetinovic, D.: IoT public fog nodes reputation system: a decentralized solution using Ethereum blockchain. IEEE Access **7**, 178082–178093 (2019)
11. Fortino, G., Fotia, L., Messina, F., Rosaci, D., Sarné, G.M.: Trust and reputation in the internet of things: state-of-the-art and research challenges. IEEE Access **8**, 60117–60125 (2020)
12. Ge, C., Ma, X., Liu, Z.: A semi-autonomous distributed blockchain-based framework for UAVs system. J. Syst. Architect. **107**, 101728 (2020)
13. Hassija, V., et al.: Fast, reliable, and secure drone communication: a comprehensive survey. IEEE Commun. Surv. Tutorials **23**(4), 2802–2832 (2021)

14. Jena, K.K., Bhoi, S.K., Behera, B.D., Panda, S., Sahu, B., Sahu, R.: A trust based false message detection model for multi-unmanned aerial vehicle network. In: 2019 Third International conference on I-SMAC (IoT in Social, Mobile, Analytics and Cloud)(I-SMAC), pp. 324–329. IEEE (2019)
15. Kapitonov, A., Lonshakov, S., Krupenkin, A., Berman, I.: Blockchain-based protocol of autonomous business activity for multi-agent systems consisting of UAVs. In: 2017 Workshop on Research, Education and Development of Unmanned Aerial Systems (RED-UAS), pp. 84–89. IEEE (2017)
16. Kong, L., Chen, B., Hu, F.: Lap-BFT: Lightweight asynchronous provable byzantine fault-tolerant consensus mechanism for UAV network. Drones 6(8), 187 (2022)
17. Lagkas, T., Argyriou, V., Bibi, S., Sarigiannidis, P.: Uav IoT framework views and challenges: towards protecting drones as "things". Sensors 18(11), 4015 (2018)
18. Makhdoom, I., Abolhasan, M., Abbas, H., Ni, W.: Blockchain's adoption in IoT: The challenges, and a way forward. J. Netw. Comput. Appl. 125, 251–279 (2019)
19. Mohsin, A.H., et al.: Blockchain authentication of network applications: Taxonomy, classification, capabilities, open challenges, motivations, recommendations and future directions. Comput. Stand. Interfaces 64, 41–60 (2019)
20. Puthal, D., Malik, N., Mohanty, S.P., Kougianos, E., Das, G.: Everything you wanted to know about the blockchain: Its promise, components, processes, and problems. IEEE Consum. Electron. Mag. 7(4), 6–14 (2018)
21. Qureshi, K.N., Jeon, G., Hassan, M.M., Hassan, M.R., Kaur, K.: Blockchain-based privacy-preserving authentication model intelligent transportation systems. IEEE Trans. Intell. Transp. Syst. 24, 7435–7443 (2022)
22. Resnick, P., Zeckhauser, R.: Trust among strangers in internet transactions: empirical analysis of Ebay's reputation system. In: The Economics of the Internet and E-commerce, vol. 11, pp. 127–157. Emerald Group Publishing Limited (2002)
23. Shehada, D., Gawanmeh, A., Yeun, C.Y., Zemerly, M.J.: Fog-based distributed trust and reputation management system for internet of things. J. King Saud Univ.-Comput. Inf. Sci. 34(10), 8637–8646 (2022)
24. Wang, Y., Wen, J., Zhou, W., Tao, B., Wu, Q., Tao, Z.: A cloud service selection method based on trust and user preference clustering. IEEE Access 7, 110279–110292 (2019). https://doi.org/10.1109/ACCESS.2019.2934153
25. Wang, Z., Xiong, R., Jin, J., Liang, C.: Airbc: a lightweight reputation-based blockchain scheme for resource-constrained UANET. In: 2022 IEEE 25th International Conference on Computer Supported Cooperative Work in Design (CSCWD), pp. 1378–1383 (2022). https://doi.org/10.1109/CSCWD54268.2022.9776299
26. Xu, R., Ramachandran, G.S., Chen, Y., Krishnamachari, B.: BlendSM-DDM: BLockchain-ENabled secure microservices for decentralized data marketplaces. In: 2019 IEEE International Smart Cities Conference (ISC2), pp. 14–17. IEEE (2019)
27. Xu, R., Wei, S., Chen, Y., Chen, G., Pham, K.: Lightman: a lightweight microchained fabric for assurance- and resilience-oriented urban air mobility networks. Drones 6(12), 421 (2022)
28. Xu, X., Zhao, H., Yao, H., Wang, S.: A blockchain-enabled energy-efficient data collection system for UAV-assisted IoT. IEEE Internet Things J. 8(4), 2431–2443 (2021). https://doi.org/10.1109/JIOT.2020.3030080
29. Yaacoub, J.P., Noura, H., Salman, O., Chehab, A.: Security analysis of drones systems: attacks, limitations, and recommendations. Internet Things 11, 100218 (2020)
30. Yu, Y., Li, Y., Tian, J., Liu, J.: Blockchain-based solutions to security and privacy issues in the internet of things. IEEE Wirel. Commun. 25(6), 12–18 (2018). https://doi.org/10.1109/MWC.2017.1800116

Resilient Range-Only Cooperative Positioning of Multiple Smart Unmanned Aerial Systems

Yajie Bao[1]([✉]), Dan Shen[1], Genshe Chen[1], Khanh Pham[2], and Erik Blasch[3]

[1] Intelligent Fusion Technology, Inc., Germantown, MD 20876, USA
{yajie.bao,dshen,gchen}@intfusiontech.com
[2] Air Force Research Lab, Kirtland AFB, Albuquerque, NM 87117, USA
khanh.pham.1@spaceforce.mil
[3] Air Force Office of Scientific Research, Arlington, VA 22203, USA

Abstract. Deploying multiple Unmanned Aerial Systems (UASs) is beneficial for applications that survey large regions and require cooperative redundancy. Range-only cooperative navigation has been proposed to enhance positioning precision by exchanging navigation information, especially in Global Navigation Satellite Systems (GNSS)-denied environments. However, existing works do not consider the possible attacks on range-only positioning in exceptionally adverse environments and do not investigate the resilience of cooperative navigation. In this paper, we consider the attacks on range measurements in the context of distributed range-only positioning using the Extended Kalman Filter (EKF) and present an anti-attack approach by integrating the Inertial Measurement Units (IMU) with the distributed position estimator. Moreover, this paper evaluates the resilience of the cooperative navigation system under Gaussian and non-Gausisian attacks. Extensive simulations on a cooperative task for multiple UASs to survey a target area demonstrate that the range-only positioning by EKF is vulnerable to non-Gaussian attacks and that the proposed anti-attack approach can detect the attacks with a high probability and mitigate the performance degradation caused by attacks.

Keywords: Resilient positioning · range-only positioning · distance manipulation attacks · cooperative positioning

1 Introduction

Positioning is an essential utility for many cyber-physical system operations such as smart vehicles and intelligent transportation. The Global Positioning Sys-

This material is partially based upon work supported by the AFRL under Contract No. FA9453-23-P-A019. The views and conclusions contained herein are those of the authors and should not be interpreted as necessarily representing the official policies or endorsements, either expressed or implied, of the United States Air Force or the U.S. Government.

© ICST Institute for Computer Sciences, Social Informatics and Telecommunications Engineering 2024
Published by Springer Nature Switzerland AG 2024. All Rights Reserved
Y. Chen et al. (Eds.): SmartSP 2023, LNICST 552, pp. 130–147, 2024.
https://doi.org/10.1007/978-3-031-51630-6_9

tem (GPS) and other Global Navigation Satellite Systems (GNSS) are accurate sources for positioning but may be vulnerable to intentional attacks [5,14,15]. There are two main types of attacks for GNSS systems: a) jamming [3] to affect the availability of the GNSS satellite signals by generating powerful signals in the GNSS band; and b) spoofing to deceive the GNSS user navigation by transmitting signals that share the same characteristics with the legitimate GNSS satellite signals [13]. GNSS spoofing can even take over the control of UASs that rely on GNSS for navigation [20]. To detect attacks, signal processing techniques based on the characterization of the attacks have been developed by checking distortions or disruptions of signals [19]. Furthermore, the integration of independent measurements and information has been considered for attack detection by monitoring drifts of the receiver position and/or clock. Moreover, simultaneously using complementary strategies has been proposed to compensate for the weaknesses of an individual attack detection technique that might be exploited by a sophisticated spoofer.

Other methods to provide security for communications include blockchain security, data encryption, user authentication, message hiding, and signal analysis. Monitoring the signals analysis can only detect spoofing and cannot correct the error [24]. A hidden message would require a larger channel capacity and methods to resolve the true signal [9]. While authentication could be a solution [26], if the signal is spoofed, it would require protocols that cause timing delays amongst many sources requiring ID-based signature message recovery [31]. Since navigation methods like GPS and automatic dependent surveillance-Broadcast (ADS-B) could add authentication, there are still ways to send incorrect messages. Encryption is challenging as it is not backward compatible and would require a fundamental alteration of the signals with standardized approaches [25]. Currently, there are efforts towards secure distributed edge-based methods [8] that could use blockchain which is popular for smart sensors [28]. Analysis of blockchain for avionics shows promise, but increases the message size, reduces timing, and requires more memory [27], and efforts are underway to make the system lighter [29]. Hence the only current solution is to have another massage source such as a designated radar signal that is typically only located at designated airports. Using another onboard edge sensing source to detect and correct the spoofing as well as be available for GNSS jamming would provide a practical solution for continuous navigation.

Range-only positioning provides an alternate source of position estimations using relative distances to fixed or dynamic beacons [2]. In the case of multiple UASs, cooperative navigation/positioning where individual UASs exchange information to improve their own position estimation has been developed for robust positioning [21]. For example, the authors [12] proposed a distributed consensus-based distributed EKF approach for collaborative relative navigation. Furthermore, observability for range-only cooperative localization using extended/unscented Kalman filters (KF) has been established [6] as well as bearings-only tracking [10]. Trajectory planning for favorable network configuration in terms of optimality measures has been studied to control the statistical properties of the localization error [17,30]. However, the existing works do not consider the attacks on the range sensors or information exchange in adverse environments which may cause large errors in range measurements besides the normal measurement noise and thus degrade the positioning performance [22]. Moreover, since the commonly used extended/unscented KF assumes Gaussian noise for range-only positioning, non-Gaussian attacks may cause severe performance degradation and escape attack detections like the innovation testing [1]. In this paper, we investigate the performance of distributed range-only positioning systems under both Gaussian and non-Gasussian attacks.

To detect and mitigate the attacks on the range-only positioning, we use inertial measurement units (IMU) as another source of positioning, similar to the integration of GNSS and IMU for anti-attacks [7,19]. It is noted that dead reckoning based on IMU measurements cannot provide precise positioning without an accurate previously determined position. However, the IMU is less susceptible to signal/data attacks. Therefore, we can combine the range-only positioning and IMU to detect attacks. By discarding the attacked UASs, the rest UASs may still achieve accurate positioning when the unattacked nodes can ensure the observability of the cooperative positioning system.

The main contribution of this paper lies in presenting a distributed EKF-based approach integrated with IMU-based positioning for the detection and mitigation of distance manipulation attacks on the range-only cooperative positioning of multiple UASs in GNSS-denied environments. The remainder of the paper is organized as follows: Sect. 2 gives the problem formulation, including the dynamic models and the preliminaries of the distributed EKF (DEKF); Sect. 3 introduces the distance manipulation attacks and anti-attack approach based on DEKF and IMU; Sect. 4 provides experimental results; and finally, Sect. 5 summarizes this paper.

2   Problem Formulation

Consider a system of multiple UASs that consist of a leader node N_0 and N_s follower nodes where $s = 1, \cdots, S$; the leader node is hovering at a position/maintains high-precision positioning while the follower nodes need to fly through a potential GPS-denied region towards a target area. Each UAS can obtain the relative distance to the leader node and the neighboring UASs using the time of arrival (TOA) mode via a data link during the flight. Moreover, the data link may be spoofed and transmit misleading range measurements. Additionally, the UASs can obtain measurements of gyro rate and acceleration from the onboard low-cost IMU, azimuth from the magnetometer, air speed measured from a Pitot tube, and height (from the ground) from a baro-altimeter.

The problem addressed in this work is how to design a resilient scheme for employing the range measurements and internal measurements to achieve an acceptable estimation of positions in the GPS-denied and/or spoofing environments, as shown by Fig. 1.

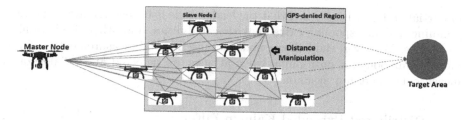

Fig. 1. Range-only positioning in GPS-denied environments under distance manipulation attacks.

2.1 Dynamic Models and Measurements

Denote x_0 as the 3-D coordinates of the leader node. Without loss of generality, we use a global coordinate/frame without considering the transition from the local frame to the global frame and $x_0 = [0, 0, 0]^\top$. The dynamic process and the local observation of each node i can be described using the following state-space model:

$$x_i(k+1) := \begin{bmatrix} x_{i,1}(k+1) & x_{i,2}(k+1) & x_{i,3}(k+1) \end{bmatrix}^\top \tag{1a}$$

$$= \begin{bmatrix} x_{i,1}(k) \\ x_{i,2}(k) \\ x_{i,3}(k) \end{bmatrix} + \begin{bmatrix} u_{i,1}(k) \\ u_{i,2}(k) \\ u_{i,3}(k), \end{bmatrix} dt + \omega_i(k), \tag{1b}$$

$$y_{i,j}(k) = \sqrt{(x_i(k) - x_j(k))^\top (x_i(k) - x_j(k))} + \nu_i(k), \tag{1c}$$

$$i = 1, \cdots, N_s, j \in \mathcal{N}_i(k), \tag{1d}$$

where x_i and u_i denote the coordinates and velocities of the i-th node, respectively; $k \in \mathbb{N}$ is the time instant and dt is the time increment; $\omega_i \in \mathbb{R}^3$ is the process noise with covariance matrix denoted by $Q(k)$; $y_{i,j}$ is the range measurements between node i and j and $\nu_i \in \mathbb{R}^3$ is measurement noise with covariance matrix denoted by $R(k)$ which is assumed to follow normal distribution; $\mathcal{N}_i$ is the set of neighboring nodes for the node i. It is noted that $\mathcal{N}_i$ is varying as a result of the dynamics of UASs. Moreover, the cardinality $|\mathcal{N}_i|$ (i.e., the number of neighbors) is a tuning parameter, which can be determined based on the verification of the measurement data for resilience.

It is noted that $y_{i,j} = y_{j,i}$ may not hold due to measurement errors. One approach for positioning is a centralized method, i.e., the follower nodes transmit the measurements to the leader and the leader uses the extended/unscented Kalman filter to estimate the positions. However, regardless of the computational and communication cost, this approach may not work in case parts of the nodes fail to transmit reliable measurements to the leader node due to interruptions of communications or spoofing.

Instead, we consider a distributed approach where each follower node uses the range measurements w.r.t. the leader node and neighboring nodes for positioning such that an acceptable estimation can still be achieved in case of failures of partial nodes. It is assumed that the leader node is far enough away, has anti-jamming and anti-spoofing extra analytic capabilities, and otherwise is resilient to attacks. Moreover, the follower node can use internal measurements and previous estimations for positioning when attacks are detected and reliable range measurements are not available.

2.2   Distributed Extended Kalman Filter

Range-only positioning requires a nonlinear state estimator due to the nonlinearity of Eq. (1c). EKF is an efficient approach for nonlinear state estimation. In particular, the EKF linearizes the nonlinear measurement and/or state transition functions using the first-order Taylor series at the current best state estimate for filtering and predictions of states. Specifically, the linearized model at time instant k is

$$x(k+1) = F(k)x(k) + G(k)\omega(k) + u(k), \tag{2a}$$

$$\bar{y}(k) \approx H(k)x(k) + \nu(k), \tag{2b}$$

where $x = [x_1^\top, \cdots, x_S^\top]^\top$ represents the augmented states that consist of the states of all the follower nodes, $F(k) = \frac{\partial f}{\partial x}|_{\hat{x}(k|k-1)}$ with $x(k+1) = f(x(k)) + G(k)\omega(k) + u(k)$ denoting the state transition function; $H(k) = \frac{\partial h}{\partial x}|_{\hat{x}(k|k-1)}$ with $y = h(x)$ denoting the nonlinear measurement functions; $\bar{y}(k) = y(k) - h(\hat{x}(k|k-1)) + H(k)\hat{x}(k|k-1)$. Then, at time instant k, the correct step based on the measurements is

$$P(k|k) = (P^{-1}(k|k-1) + H^\top(k)R^{-1}(k)H(k))^{-1}, \tag{3a}$$

$$\hat{x}(k|k) = \hat{x}(k|k-1) + P(k|k)H^\top(k)R^{-1}(k)\left(\bar{y}(k) - H(k)\hat{x}(k|k-1)\right); \tag{3b}$$

the prediction step is

$$\hat{x}(k+1|k) = f(\hat{x}(k|k)), \tag{4a}$$

$$P(k+1|k) = F(k)P(k|k)F^\top(k) + G(k)Q(k)G^\top(k). \tag{4b}$$

Instead, the distributed EKF uses the local measurements for correction and prediction and obtains an accurate estimate of the entire system state variables based on consensus [4]. In particular, the consensus-based correct step [23] for the i-th node in a network of homogeneous nodes is

$$\hat{x}_i^l(k|k) =$$

$$\left(\frac{1}{N_s^+} P_i^{-1}(k|k-1) + H_i^\top(k) R_i^{-1}(k) H_i(k) \frac{|\mathcal{N}_i^+|}{\mu} I \right)^{-1} \cdot$$

$$\left[H_i^\top(k) R_i^{-1}(k) \bar{y}_i(k) + \frac{1}{N_s^+} P_i^{-1}(k|k-1) \hat{x}_i(k|k-1) + \right.$$

$$\left. \sum_{j \in \mathcal{N}_i^+} \left(\frac{z_j(k)^{l-1}}{\mu} + \lambda_{ij}^{l-1} \right) \right] \tag{5a}$$

$$z_i(k)^l = \frac{\mu}{|\mathcal{N}_i^+|} \sum_{j \in \mathcal{N}_i^+} \left(\frac{1}{\mu} \hat{x}_j^l(k|k) - \lambda_{ji}^{l-1} \right), \tag{5b}$$

$$\lambda_{ij}^l = \lambda_{ij}^{l-1} - \frac{1}{\mu} \left(\hat{x}_i^l(k) - z_j^l(k) \right), \tag{5c}$$

$$\forall\, i = 1, \cdots, N_s,\; j \in \mathcal{N}_i^+,\; l = 1, \cdots, L \tag{5d}$$

where $\hat{x}_i^l$ is Node i's estimate of x using local P_i, H_i, and R_i at Node i for the l-th iteration, and $P_i(0|0) = P_0$; z_i^l is the auxiliary variable with initialization $z_i^0(k) = \hat{x}_i(k|k-1)$, $\lambda_{i,j}^l$ is the Lagrange multiplier with initialization $\lambda_{i,j}^0 = 0$, and μ is a scalar penalty parameter; $z_j, \hat{x}_j, j \in \mathcal{N}_i$ are transmitted from the $|\mathcal{N}_j|$ nearest neighbors of Node i based on the noisy range measurements; $N_s^+ = N_s + 1$, and $\mathcal{N}_i^+ = \mathcal{N}_i \cup \{i\}$; the correction of the covariance matrix is

$$P_i(k|k) = \left(P_i^{-1}(k|k-1) + \sum_{i=1}^{N_s} H_i^\top(k) R_i^{-1}(k) H_i(k) \right)^{-1}; \tag{6}$$

The prediction step for the i-th node is

$$\hat{x}_i(k+1|k) = f(\hat{x}_i(k|k)), \tag{7a}$$

$$P_i(k+1|k) = F_i(k) P_i(k|k) F_i^\top(k) + G(k) Q(k) G^\top(k). \tag{7b}$$

The main advantage of the DEKF approach is that it can reduce the computational burden and communication overhead as compared to a centralized approach. The DEKF can be more scalable and robust versus a centralized (CEKF), especially in systems with a large number of sensors distributed across different locations and limited, unreliable, or costly communication between nodes.

3   Distance Manipulation Attacks

In this section, we introduce the distance manipulation attacks on the range-only cooperative positioning and present the proposed approaches to detecting and preventing the attacks.

3.1   Attacks on Range Measurements

The demand for ranging information is increasing for autonomous and cyber-physical systems in various applications such as positioning and navigation, which makes it a target of attackers with different motivations. Existing ranging systems such as ultra-wideband (UWB) ranging systems are vulnerable to distance manipulation attacks. Distance manipulation attacks can be performed by manipulating the logical or physical layer. Logical-layer attacks manipulate message bits while physical-layer attacks involve manipulating signal characteristics to incorrectly measure the signal's phase, amplitude, frequency, or arrival time [22]. Additionally, distance manipulation attacks can be divided into distance reduction and enlargement attacks. An attacker may reduce the measured distance by manipulating the time of arrival (ToA) estimation of the preamble (via cicada attack [18]) and the payload (via Early Detect Late Commit (ED/LC) attack) [11] and enlarge the measured distance by preventing legitimate payload detection by increasing the bit error by adding noise in the channel or canceling some of the pulses. The availability of affordable radio devices like the software-defined radio has opened up vast possibilities for cybersecurity and infosec professionals to explore radio frequency (RF) communication and control devices, enabling them to delve into hacking in this domain.

In the case of range-only positioning, we consider the distance manipulation attacks introducing extra range measurement disturbances. Specifically, the attacked range measurements

$$\tilde{y}_{i,j}(k) = \begin{cases} y_{i,j}(k) + b_{i,j}(k), & i \in \mathcal{A}_V(k), k \in \mathcal{A}_T \\ y_{i,j}(k) & \text{otherwise} \end{cases}, \qquad (8)$$

where $b_{i,j}(k)$ denotes the modification of Node i's measurement of the range between Node i and j at time k; $\mathcal{A}_V(k)$ is the set of attacked nodes and $\mathcal{A}_T$ is the set of attacked time steps. Then, the centralized/distributed EKF use $\tilde{y}_{i,j}(k)$ at each time step to correct positioning estimation, which may cause large deviations from the real positioning.

3.2   Attack Detection and Mitigation

Using alternative positioning sources is a common strategy to detect and mitigate attacks. In addition to the range measurements, IMU measurements can be used for positioning. In particular, the raw IMU measurements can be utilized to calculate position relative to a global reference frame via a method known as dead reckoning. Using a previously determined position, dead reckoning can provide an accurate current position by $\hat{x}(k) = \hat{x}(k-1) + \delta\hat{x}(k-1)$ where $\delta\hat{x}$ denotes the displacement computed by the data of IMU sensors. Moreover, the IMU is less vulnerable to attacks than range-only positioning for which communications between nodes are required. However, dead reckoning is subject to cumulative errors over time and causes significant drifts over great distances.

For attack detection of GNSS and IMU, innovation testing [1] is widely used. However, the EKF may mitigate the effects of attacks such that the differences between the position estimates of range and IMU measurements are unreliable for detecting attacks on range measurements. In consequence, accumulating the faults within a time window is needed to detect the slowly drifting faults introduced by GNSS spoofing attacks [16], which may disable the in-time detection and mitigation of attacks. Instead, we use the differences between the range measurements and the range estimates based on the IMU measurements to detect the attacks, as the dead reckoning can maintain high accuracy for a short period and is less vulnerable to communication attacks. In particular,

$$\mathbb{1}_i^a(k) = \begin{cases} 1, & \text{if } \exists j \text{ s.t. } |\hat{y}_{i,j}(k) - y_{i,j}(k)| > \gamma \\ 0 & \text{otherwise} \end{cases} \tag{9}$$

where $\mathbb{1}_i^a$ indicates whether the i-th node is attacked and γ is a predefined threshold and $\hat{y}_{i,j}(k)$ is the range estimates between Node i and Node j based on the IMU measurements. It is noted that there can be detection errors including false alarms and mis-detections.

Then, we combine range-based positioning and dead reckoning of IMU to enhance the resilience of the positioning system. In particular, we use the range-only positioning in the normal environment and IMU when attacks occur. To avoid the drifts of IMU-based positioning, the IMU is calibrated using the range-only positioning at a predefined frequency, when no attacks are detected. However, the IMU will not be calibrated once the attacks are detected, and dead reckoning will be used for positioning until the attack alarms cease. The difference between the range measurements and estimates will be monitored in real-time to detect attacks.

Furthermore, we consider two cases of reducing performance degradation when the attacks are detected. First, when the number of unattacked nodes based on the detection is greater than the number of nodes required for distributed EKF, the information from the attacked nodes will be discarded to prevent the adverse effects of incorrect measurements. The second case is when the number of unattacked nodes is less than the number of nodes required for DEKF, the attacked nodes use the IMU-based position estimates as their position estimates. Additionally, the procedures for attack detection and mitigation are summarized in Algorithm 1.

4   Experimental Results and Validation

4.1   Scenario Description

Table 1. Specifications of the UVA

Cruising Speed	Range	Endurance	Height	Field of View
30 km/h	10 km	1–1.5 h	0.15 km	31.5°–6.7°

Algorithm 1. Detecting and Mitigating Distance Manipulation Attacks

---

Input: S: number of follower nodes; $x(0)$: initial positions; x_{target}: target coordinates; ϵ: target radius; $\hat{P}_i(0)$, initial estimate of state covariance matrix; Q: process noise covariance matrix; R: measurement noise covariance matrix; $|\mathcal{N}_i|$: the number of neighboring nodes for Node i; N_{DEKF}: the number of neighboring nodes required for DEKF; T: maximum time step.

Output: $x_i(k), i = 1, \cdots, S, k = 0, \cdots, T$.

1: **Initialization:** $k = 0$, $\mathbb{1}_{\text{IMU}} = \text{False}$ ▷ $\mathbb{1}_{\text{IMU}} = \text{False}$ if calibrating IMU with the range-only positioning, and $\mathbb{1}_{\text{IMU}} = \text{True}$ otherwise.
2: **while** $\max_i \|x_i(k) - x_{\text{target}}\| > \epsilon$ and $k < T$ **do**
3:     Initialize $\mathcal{N}^a(k) = \varnothing$ at time instant k ▷ $\mathcal{N}^a(k)$: the set of attacked nodes.
4:     **for** $i \leftarrow 1$ to S **do**
5:         Compute and apply control input $u_i(k)$ based on $\hat{x}_i(k)$
6:         Obtain range and IMU measurements
7:         **if** $\mathbb{1}_{\text{IMU}}$ **then**
8:             Obtain IMU-based position estimates $\hat{x}_i^{\text{IMU}}(k)$ using $\hat{x}_i^{\text{IMU}}(k-1)$
9:         **else**
10:            Obtain IMU-based position estimates $\hat{x}_i^{\text{IMU}}(k)$ using $\hat{x}_i(k-1)$
11:        **end if**
12:        Obtaining IMU-based distance estimates based on $\hat{x}_i^{\text{IMU}}(k)$
13:        **if** $\mathbb{1}_i^a$ **then**   ▷ Attack detection by Eq. (9).
14:            $\mathcal{N}_a(k) = \mathcal{N}_a(k) \cup \{i\}$
15:            $\mathbb{1}_{\text{IMU}} = \text{True}$
16:        **else**
17:            $\mathbb{1}_{\text{IMU}} = \text{False}$
18:        **end if**
19:    **end for**
20:    **if** $|\mathcal{N}^{\bar{a}}(k)| \geq N_{\text{DEKF}}$ **then**   ▷ $|\mathcal{N}^{\bar{a}}(k)|$: the number of unattacked nodes.
21:        **for** $i \leftarrow 1$ to S **do**
22:            Estimate $\hat{x}_i(k)$ using the DEKF with $\mathcal{N}_i^{\bar{a}} \cap \mathcal{N}_i$
23:        **end for**
24:    **else**
25:        **for** $i \leftarrow 1$ to S **do**
26:            $\hat{x}_i(k) = \hat{x}_i^{\text{IMU}}(k)$
27:        **end for**
28:    **end if**
29:    $k \leftarrow k + 1$
30: **end while**

We assume each UAS to be a point UAS and that there are no kinematic restrictions on a UAS's movement, similar to [17]. The UAV specifications[1] are summarized in Table 1. The leader node stays at $x_0 = [0\ 0\ 0]^\top$ (m). The initial positions of the follower nodes $x_i(0) = [x_{i,1}(0)\ x_{i,2}(0)\ x_{i,3}(0)]^\top + [0\ 0\ 150]^\top$ where $x_{i,j}(0)$ are randomly drawn from the normal distribution $\mathcal{N}(0, 0.1)$. The target is $x_{\text{target}} = [5000\ 5000\ 150]^\top$ (m). There are S^2 range measurements, including

---

[1] We refer to Raven® B RQ-11 at https://www.avinc.com/images/uploads/product_docs/Raven_Datasheet_05_220825.pdf for the specifications.

the measurements between the leader node and follower nodes and between each two follower nodes. Each node adjusts the control inputs u_i by

$$u_i(k) = \frac{\hat{x}_i(k|k) - x_{\text{target}}}{\|\hat{x}_i(k|k) - x_{\text{target}}\|_2} \times 8 \text{ (m/s)}, \tag{10}$$

where $\hat{x}_i(k|k)$ is the position estimate at time k based on the measurements and $\|\cdot\|_2$ denotes the Euclidean norm. $dt = 9$ s. The covariance matrix R of the measurement noises is diagonal and $R = \text{diag}([\sigma_{1,0}, \cdots, \sigma_{S,S-1}])$ where $\sigma_{i,j}$ denotes the standard deviation of the noise for node i's range measurement w.r.t. node j. A follower node finishes the task if $\|x_i(k) - x_{\text{target}}\| \leq 16$ (m). The maximal time steps for the task is 100. Additionally, the process noise is not considered.

Moreover, spoofing can take place on the data during the flights. To thoroughly test the performance of the range-only cooperative positioning and anti-attack techniques under various types of attacks, we consider both (1) non-Gaussian attacks which add a fixed y_a to the measurements of the $|\mathcal{A}_V|$ attacked follower nodes with a probability p_a during the attack period from time step 21 to 30; and (2) Gaussian attacks which add i.i.d. Gaussian noise with $y_a \sim \mathcal{N}(0, \sigma_a)$ to the measurements of the $|\mathcal{A}_V|$ attacked follower nodes during the attack period. The non-Gaussian attacks are supposed to cause more performance degradation and bring more challenges for anti-attacks than the Gaussian attacks, as the EKF assumes Gaussian process and measurement noise. Furthermore, since we assume homogeneous follower UAV nodes, the attacked UAV nodes are randomly selected given a number of attacked nodes.

We use a measurement-level simulator which is sufficient for attack detection and impact moderation of spoofing. To evaluate the range-only positioning approach, we use the average estimation errors computed by

$$\bar{e} = \frac{1}{M} \sum_{l=1}^{M} \frac{1}{S} \sum_{i=1}^{S} \frac{1}{K} \sum_{k=1}^{K} \left\| \hat{x}_i^{(l)}(k|k) - x_i^{(l)}(k) \right\|_2, \tag{11}$$

where M is the number of Monte Carlo (MC) simulations, and K is the number of time steps for the i-th node. Moreover, we evaluate the *success rate* which is defined as the ratio of the number of follower nodes that reach the target area over the total number of follower nodes in a simulation, and the average success rate is the average of the success rates of M MC simulations.

4.2 Performance of Centralized EKF

The Centralized EKF (CEKF) requires the follower nodes to send their range measurements to the leader node to estimate the positions of all the nodes. Then, the leader node sends the position estimates to the follower nodes. In the experiments, we assume the timing is synchronized for all the nodes and omit the processing and communicating time to focus on the positioning problem. First, we evaluate the performance of CEKF for different numbers of follower nodes and different σ_ν's.

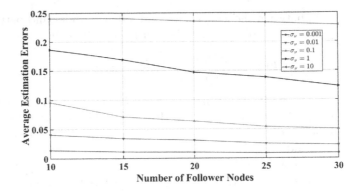

(a) Average estimation errors of CEKF.

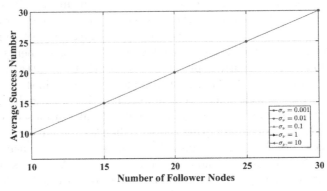

(b) Average success number of CEKF.

Fig. 2. The performance of CEKF for different numbers of follower nodes.

Results. Figure 2 shows the results of evaluating CEKF. The average estimation errors decrease as the number of follower nodes increases and increase as the standard deviation of the measurement noise increases. All the follower nodes fulfilled the tasks for different measurement noises when no attacks took place, which demonstrates the good performance of CEKF when its assumptions are satisfied.

4.3   Performance of Distributed EKF

The distributed EKF (DEKF) requires each follower node to estimate positions based on the consensus with the neighboring follower nodes. In particular, Node i needs to update and transmit $\hat{x}_i^l(k|k)$ and $z_i^l(k|k), l = 1, \cdots, L$ for consensus. We consider $S = 10$ for the following simulations. The parameters for consensus in (5) are determined as $L = 40, \mu = 0.1$. We assume all the follower nodes can maintain the range measurements w.r.t. the leader node. First, we evaluate the performance of the DEKF using different numbers of neighboring nodes.

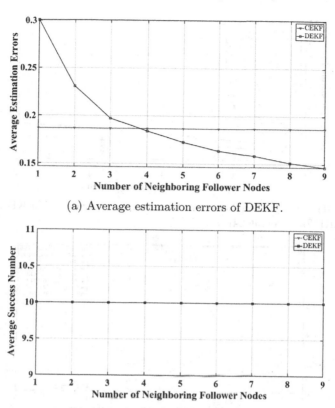

(a) Average estimation errors of DEKF.

(b) Average success number of DEKF.

Fig. 3. The performance of DEKF for different numbers of neighboring nodes.

Results. Figure 3 shows the results of evaluating DEKF. The average estimation errors decrease as the number of neighboring follower nodes increases and are smaller than those of CEKF when the number of neighboring nodes is greater than 3. All the follower nodes fulfilled the tasks for different measurement noises when no attacks took place, which demonstrates that DEKF can achieve acceptable precision without using all the measurements as CEKF.

4.4 Resilience Against Attacks

In this subsection, we evaluate the performance of range-only positioning under attacks. First, we evaluate the performance degradation under different realizations of attacks and show the performance of dead reckoning using IMU. Then, we validate the proposed attack detection and mitigation approach. Additionally, the number of neighboring follower nodes for the experiments in this subsection was set to 4 which is sufficient for DEKF to achieve comparable precision with CEKF based on the results in Sect. 4.3 (Fig. 4).

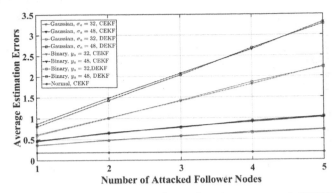

(a) The comparison between the estimation errors of CEKF and DEKF for different $|\mathcal{A}_V|$.

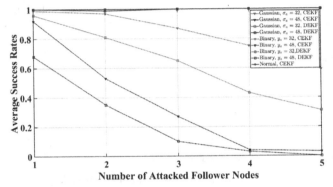

(b) The comparison between the average success rates of CEKF and DEKF for different $|\mathcal{A}_V|$.

Fig. 4. The comparison between the performance of CEKF and DEKF under attacks.

Performance Under Attacks. As the number of attacked follower nodes increases, the average estimation errors increase. Non-Gaussian attacks cause more significant decreases in performance. The differences between the average estimation errors of CEKF and DEKF were not significant. However, CEKF achieved higher average success rates than DEKF without anti-attack techniques.

Performance of Dead Reckoning. Since the considered attacks only impact the range measurements, the IMU will not be affected but still suffer from the drifts by dead reckoning. For simulations, we assume that the velocity estimated from the IMU measurements suffers from an additive Gaussian noise with $\sigma_{IMU} = 0.1(m/s)$. The red line with downward-pointing triangle marks in Fig. 5 shows the average estimation errors over time using only IMU in one simulation. The average estimation errors at a time step are the average of the estimation

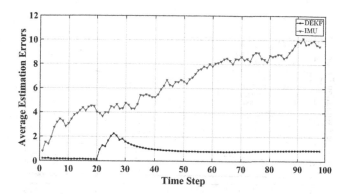

Fig. 5. The average estimation errors over time using only IMU for positioning.

errors of the 10 follower nodes at that time step. Additionally, the black line with the square marks shows the result of DEKF without using the anti-attack technique in one simulation under non-Gaussian attacks with $y_a = 48$ and $|\mathcal{A}_V| = 5$. 10 follower nodes finished the task using DEKF in that simulation.

Performance of the Anti-attack Approach. Figure 6 shows the validation results of the anti-attack approach. We evaluated the detection errors of the proposed attack detection approach. The detection error is the number of false detections (including false positives and false negatives) in a simulation and the average detection errors are the average of the detection errors of M simulations. The average detection errors for Gaussian attacks were larger than those of non-Gaussian attacks. The proposed detection approach correctly detected the attacked nodes with a high probability (that is greater than 0.94 for all the attacks), selected the unattacked neighboring nodes for consensus-based DEKF, and achieved similar performance to the CEKF without attacks (black lines with square marks).

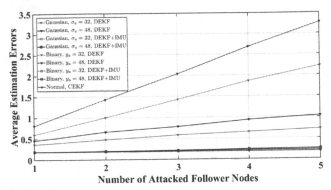

(a) The comparison between the estimation errors of DEKF without and with anti-attack for different $|\mathcal{A}_V|$.

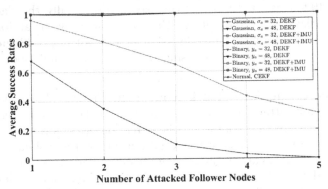

(b) The comparison between the average success rates of DEKF without and with anti-attack for different $|\mathcal{A}_V|$.

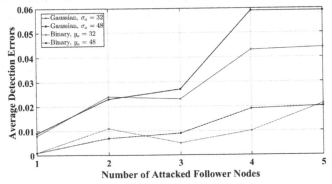

(c) The detection errors for different $|\mathcal{A}_V|$, attack types, and attack strength.

Fig. 6. Validation results of the anti-attack approach.

5   Concluding Remarks

This paper presented a DEKF approach to detecting and mitigating distance manipulation attacks on range-only positioning of multiple smart UASs with IMU-based positioning. In particular, both Gaussian and non-Gaussian types of attacks were considered. The attacks were detected based on the differences between the IMU-based distance estimates and the range measurements and the UAS exchanged information with adjacent UASs that were free of attacks to enhance positioning precision. Experiments demonstrated that DEKF were more robust to attacks than CEKF and using the anti-attack approach based on DEKF and IMU further reduced the positioning errors and improved the probability of fulfilling tasks.

In future works, we will consider other types of distance manipulation attacks and enhance the attack detection and mitigation performance by statistical analysis and machine learning algorithms.

References

1. Anderson, B.D., Moore, J.B.: Optimal Filtering. Courier Corporation, Chelmsford (2012)
2. Bahr, A., Leonard, J.J., Martinoli, A.: Dynamic positioning of beacon vehicles for cooperative underwater navigation. In: 2012 IEEE/RSJ International Conference on Intelligent Robots and Systems, pp. 3760–3767. IEEE (2012)
3. Bao, Y., et al.: PID-based automatic gain control for satellite transponder under partial-time partial-band AWGN jamming. In: Sensors and Systems for Space Applications XVI, vol. 12546, pp. 61–68. SPIE (2023)
4. Battistelli, G., Chisci, L., Mugnai, G., Farina, A., Graziano, A.: Consensus-based linear and nonlinear filtering. IEEE Trans. Autom. Control **60**(5), 1410–1415 (2014)
5. Blasch, E., et al.: Cyber awareness trends in avionics. In: 2019 IEEE/AIAA 38th Digital Avionics Systems Conference (DASC), pp. 1–8. IEEE (2019)
6. Burchett, B.T.: Unscented Kalman filters for range-only cooperative localization of swarms of munitions in three-dimensional flight. Aerosp. Sci. Technol. **85**, 259–269 (2019)
7. Ceccato, M., Formaggio, F., Laurenti, N., Tomasin, S.: Generalized likelihood ratio test for GNSS spoofing detection in devices with IMU. IEEE Trans. Inf. Forensics Secur. **16**, 3496–3509 (2021)
8. Chen, N., Chen, Y., Blasch, E., Ling, H., You, Y., Ye, X.: Enabling smart urban surveillance at the edge. In: 2017 IEEE International Conference on Smart Cloud (SmartCloud), pp. 109–119. IEEE (2017)
9. Cheng, X.J., Xu, J.n., Cao, K.J., Wang, J.: An authenticity verification scheme based on hidden messages for current civilian GPS signals. In: 2009 Fourth International Conference on Computer Sciences and Convergence Information Technology, pp. 345–352. IEEE (2009)
10. Dunik, J., Straka, O., Simandl, M., Blasch, E.: Random-point-based filters: analysis and comparison in target tracking. IEEE Trans. Aerosp. Electron. Syst. **51**(2), 1403–1421 (2015)

11. Flury, M., Poturalski, M., Papadimitratos, P., Hubaux, J.P., Le Boudec, J.Y.: Effectiveness of distance-decreasing attacks against impulse radio ranging. In: Proceedings of the Third ACM Conference on Wireless Network Security, pp. 117–128 (2010)
12. Gong, B., Wang, S., Hao, M., Guan, X., Li, S.: Range-based collaborative relative navigation for multiple unmanned aerial vehicles using consensus extended kalman filter. Aerosp. Sci. Technol. **112**, 106647 (2021)
13. Ioannides, R.T., Pany, T., Gibbons, G.: Known vulnerabilities of global navigation satellite systems, status, and potential mitigation techniques. Proc. IEEE **104**(6), 1174–1194 (2016)
14. Kassas, Z.M., Closas, P., Gross, J.: Navigation systems panel report navigation systems for autonomous and semi-autonomous vehicles: current trends and future challenges. IEEE Aerosp. Electron. Syst. Mag. **34**(5) (2019)
15. Kassas, Z.M., Khalife, J., Abdallah, A.A., Lee, C.: I am not afraid of the Gps jammer: resilient navigation via signals of opportunity in GPS-denied environments. IEEE Aerosp. Electron. Syst. Mag. **37**(7), 4–19 (2022)
16. Liu, Y., Li, S., Fu, Q., Liu, Z., Zhou, Q.: Analysis of Kalman filter innovation-based GNSS spoofing detection method for INS/GNSS integrated navigation system. IEEE Sens. J. **19**(13), 5167–5178 (2019)
17. Papalia, A., Thumma, N., Leonard, J.: Prioritized planning for cooperative range-only localization in multi-robot networks. In: 2022 International Conference on Robotics and Automation (ICRA), pp. 10753–10759. IEEE (2022)
18. Poturalski, M., Flury, M., Papadimitratos, P., Hubaux, J.P., Le Boudec, J.Y.: The cicada attack: degradation and denial of service in IR ranging. In: 2010 IEEE International Conference on Ultra-Wideband, vol. 2, pp. 1–4. IEEE (2010)
19. Psiaki, M.L., Humphreys, T.E.: GNSS spoofing and detection. Proc. IEEE **104**(6), 1258–1270 (2016)
20. Sathaye, H., Strohmeier, M., Lenders, V., Ranganathan, A.: An experimental study of GPS spoofing and takeover attacks on UAVs. In: 31st USENIX Security Symposium (USENIX Security 22), pp. 3503–3520 (2022)
21. Shen, D., Chen, G., Cruz, J.B., Blasch, E.: A game theoretic data fusion aided path planning approach for cooperative UAV ISR. In: 2008 IEEE Aerospace Conference, pp. 1–9. IEEE (2008)
22. Singh, M.: Securing distance measurement against physical layer attacks. Ph.D. thesis, ETH Zurich (2021)
23. Wang, S., Dekorsy, A.: Distributed consensus-based extended Kalman filtering: a Bayesian perspective. In: 2019 27th European Signal Processing Conference (EUSIPCO), pp. 1–5. IEEE (2019)
24. Wen, H., Huang, P.Y.R., Dyer, J., Archinal, A., Fagan, J.: Countermeasures for GPS signal spoofing. In: Proceedings of the 18th International Technical Meeting of the Satellite Division of the Institute of Navigation (ION GNSS 2005), pp. 1285–1290 (2005)
25. Wesson, K., Rothlisberger, M., Humphreys, T.: Practical cryptographic civil GPS signal authentication. NAVIGATION: J. Inst. Navig. **59**(3), 177–193 (2012)
26. Wesson, K.D., Rothlisberger, M.P., Humphreys, T.E.: A proposed navigation message authentication implementation for civil gps anti-spoofing. In: Proceedings of the 24th International Technical Meeting of the Satellite Division of the Institute of Navigation (ION GNSS 2011), pp. 3129–3140 (2011)

27. Xu, R., Chen, Y., Blasch, E., Aved, A., Chen, G., Shen, D.: Hybrid blockchain-enabled secure microservices fabric for decentralized multi-domain avionics systems. In: Sensors and Systems for Space Applications XIII, vol. 11422, pp. 150–164. SPIE (2020)
28. Xu, R., Nikouei, S.Y., Chen, Y., Blasch, E., Aved, A.: Blendmas: a blockchain-enabled decentralized microservices architecture for smart public safety. In: 2019 IEEE International Conference on Blockchain (Blockchain), pp. 564–571. IEEE (2019)
29. Xu, R., Wei, S., Chen, Y., Chen, G., Pham, K.: Lightman: a lightweight microchained fabric for assurance-and resilience-oriented urban air mobility networks. Drones 6(12), 421 (2022)
30. Yang, C., Kaplan, L., Blasch, E., Bakich, M.: Optimal placement of heterogeneous sensors for targets with Gaussian priors. IEEE Trans. Aerosp. Electron. Syst. 49(3), 1637–1653 (2013)
31. Yang, H., Huang, R., Wang, X., Deng, J., Chen, R.: EBAA: an efficient broadcast authentication scheme for ads-b communication based on IBS-MR. Chin. J. Aeronaut. 27(3), 688–696 (2014)

Securing the Future: Exploring Privacy Risks and Security Questions in Robotic Systems

Diba Afroze⬤, Yazhou Tu⬤, and Xiali Hei⁽✉⁾⬤

University of Louisiana at Lafayette, Lafayette, LA 70504, USA
`xiali.hei@louisiana.edu`

Abstract. The integration of artificial intelligence, especially large language models in robotics, has led to rapid advancements in the field. We are now observing an unprecedented surge in the use of robots in our daily lives. The development and continual improvements of robots are moving at an astonishing pace. Although these remarkable improvements facilitate and enhance our lives, several security and privacy concerns have not been resolved yet. Therefore, it has become crucial to address the privacy and security threats of robotic systems while improving our experiences. In this paper, we aim to present existing applications and threats of robotics, anticipated future evolution, and the security and privacy issues they may imply. We present a series of open questions for researchers and practitioners to explore further.

Keywords: Robotics · Security · Privacy · Artificial Intelligence · Autonomous Device · Risk Analysis

1 Introduction

The twenty-first century is witnessing an unprecedented increase in the evolution and utilization of robots. With the upcoming Industry 4.0 revolution, we are approaching the era of robotics [39]. Currently, robotic systems play an important role, from performing medical procedures to serving as salespeople in shopping centers. Robots are now even replacing human companions. This remarkable growth, from a simple machine to an autonomous humanoid robot, has become possible because of the advancement of Artificial Intelligence, Natural Language Processing, Sensor Technology, and Processing Power.

To employ automation in work, different types of robots are used, designed to suit the specific nature of the work. We can categorize three general types of robots, i.e., *Industrial Robots*, *Service Robots*, and *Specialized Robots* [23]. Nowadays, these robots perform multipurpose applications seamlessly alongside humans in industries as well as at home. They handle heavy, mundane tasks for humans effortlessly. Additionally, they are becoming reliable in specialized tasks like healthcare assistance, surveillance, space exploration, rescue missions, etc.

© ICST Institute for Computer Sciences, Social Informatics and Telecommunications Engineering 2024
Published by Springer Nature Switzerland AG 2024. All Rights Reserved
Y. Chen et al. (Eds.): SmartSP 2023, LNICST 552, pp. 148–157, 2024.
https://doi.org/10.1007/978-3-031-51630-6_10

Robots are also helping as nurses or companions for older people. The vehicle industry is being revolutionized by the uprising of autonomous vehicles. All these advancements illustrate the prospect of reducing the gap between science fiction and reality.

As we embrace the help of robots in our daily lives, it may not be very long before these intelligent machines start to co-exist with us in society in every sector. Robotic help can undoubtedly simplify our lives, but it comes with potential privacy and security risks to our personal and social lives. Therefore, it is imperative to develop methods to prevent different kinds of privacy and security threats of robots to humans. Existing versions of robots are not free from threats, thereby indicating that future versions are unlikely to be different. There are several questions concerning privacy and security that a robot must answer before we may consider it to be safe to release in society. If we do not ensure that robots' mechanisms can answer these questions, we might have to reassess the deployment of robot among humans due to the inherent risk it poses to human life. In this paper, we explore a few of these questions.

In the following sections of this paper, we will address the growth of robotic advancement and several privacy and security-related questions that need our attention.

2   Literature Review

The proliferation of Robots is accelerating rapidly in our daily lives, and with it comes a rise in potential dangers. From the beginning of the use of robots, back in 1979, the first death induced by an industrial robot has been recorded [53]. After that, several deaths and injuries were caused by robots [25]. Even though robot R&D companies are trying to implement policies for secure interaction between humans and robots, new threats arise with the development of new robot technologies.

Today, Robots are serving in many roles, such as security guards, salespeople, helping hands at home, nurses, etc. In emergency situations, humans might not follow the instructions of robots acting as security guards [2]. An open question is: What would happen if people refused to take commands from robots? Will the robot force humans or let them pass? Trust has not yet been fully established for robot services. People are concerned about their security; They are skeptical about letting unknown robots into their living spaces [8]. Trust also depends on the appearance of robots; in some cases, people may feel threatened by humanoid robots that perform better than them at work [57].

Robots are vulnerable to various forms of cyberattacks. Clark et al. present different cyber attack scenarios [11], for example, buffer overflow attacks to take control over companion robots, attacks on automated vehicles during firmware updates by pushing corrupted updates, hardware backdoor attacks on military drones, etc. Additionally, researchers show a comprehensive view of several cybersecurity issues such as malware, Trojan, replay attacks, fault injection, tampering attacks, etc. [28,54,58].

Automated vehicles can be one of the targets of attackers. The attackers may use jamming, high-brightness Infrared LEDs, Digital Radio Frequency Memory (DRFM), etc. [40], to provide false navigation data. Additionally, autonomous vehicles are generally connected to users' smartphones. Sugawara et al. [46] presented an audio injection attack on the voice-controlled smartphone system connected to automated Tesla and Ford cars. In addition, the classification system of autonomous vehicles is at risk of potential attack. The work in [15,31] demonstrated that a simple perturbation of the traffic signal could make the CNN classification model misidentify the signal. This attack poses significant security risks and can potentially cause chaos on roadways. Unmanned Automated Vehicles (UAVs), such as drones and rovers, are also in danger of being attacked. Dash et al. [13] demonstrated three attacks on UAVs protected by control invariants (CI) [10] and the extended Kalman filter (EKF) [9]. The authors designed the attacks on UAVs by injecting minor false data into the control system, which caused the automated vehicle to change its position and angular orientations, injecting time delays to make the UAV receive commands late, and lastly, injecting malicious code to switch the mode of the UAVs. In [50], Tu et al. presented two attacks (i.e., Side Swing [22], and DoS [21]) to cyber-physical systems, and they manipulated two automatic self-balancing robots by spoofing embedded Micro Electro Mechanical Systems (MEMS) inertial sensors.

Telerobots [38] come in handy in medical surgery, military operations, and rescue missions. In [5,7], the authors elaborated that telerobots are vulnerable to common cyber attacks such as viruses, worms, and malware. They also mention security threats such as command manipulation, denial of service, and communication loss. Recently, several medical centers have filed lawsuits against Intuitive Surgical, a surgical robot manufacturer, alleging that they were coerced into signing restrictive repair contracts, forcing them to buy new parts from the aforementioned company [42]. An operation had to be postponed due to the usage of third-party repair. This incident adds another dimension to the challenges of surgical robots. Shah et al. [44] demonstrated a successful side-channel attack-*Fingeprint* on surgical robots. Besides, other potential side-channel attacks on robots are Radio-frequency attacks [45] and cache-based attacks on automated vehicles [32].

Lutz et al. [33] observed robot usage from a different perspective, implying that social robots might affect the psychological and social privacy of human beings. Van et al. [17] express their concern about whether we are compromising privacy in exchange for robotic services. *The Guardian* reported [18] about wifi-enabled Barbie dolls, which can be hacked and turned into a surveillance device to spy and collect information without anyone's knowledge. Robots are also becoming companions of humans, sometimes as caregivers. However, some authors are concerned about ethical issues. For example, the authors fear that companion robots might create a hallucinatory reality for some people [6].

3   Future Evolution and Security Questions

Robots are evolving and becoming more intelligent, precise, and *human-like*. Understandably, people are apprehensive about whether robots are going to be a threat to our lives, as depicted in science fiction movies. We are going to elaborate on some sectors for possible futuristic advancements in robots and the privacy and security questions that come with them.

- **Cyber Security:** Robots are now connected to wired and wireless networks for smooth data exchange and communication like any other device. However, robots have a lot of security issues, such as lack of authorization, authentication, secure network, tamper-resistant hardware, privacy, integrity, etc. [54]. Robotic networks and computer networks are different in nature; the same countermeasures in general computers may not work on robotics networks [52]. Robotic Operating System (ROS) is also becoming popular among developers. Nevertheless, ROS is vulnerable to attacks such as DoS, DDoS attacks, malware, buffer overflow, malicious code injection attacks, etc. [11].

  Ransomware is another concern for robot users. In [34], Mayoral-Vilches et al. show a ransomware attack-*Akerbeltz* on industrial robots, which locks and encrypts the robot from its vendor network. The attack was carried out by simply connecting a USB device to the robot or remotely accessing the adjacent network. Furthermore, another ransomware attack was demonstrated on a SoftBank Robotics NAO humanoid robot [29].

  > **Open Question 1:** Is there a way to identify security vulnerabilities early in robots? Is the robotic system software updated, or are security patches issued promptly?

- **IoT Connections:** Robots are now becoming part of IoT and interconnecting with other devices. In homes, industries, and offices, it is common to connect robots with home assistants, smartphones, and TVs. Consider a scenario where an industrial robot integrates with other devices within a multi-purpose company. If an unauthorized user takes control of the robot, the whole system will be compromised. The attacker can take control of other devices and perform dangerous tasks. For example, this security breach may lead to injury, financial damage, and data theft. Thus, it is necessary to secure the additional mobile attack interface - robots. Another scenario is depicted by Amoozadeh et al. [4], where each vehicle receives beacon messages from the immediately preceding vehicle using the IEEE 802.11p protocol. The authors demonstrated security (e.g., message falsification attack, spoofing attack, distributed DoS, Radio jamming, etc.), system-level attacks (e.g., hardware or software tempering), and privacy attacks (e.g., eavesdropping attack) on different layers of automated vehicle networks. A compromised network of vehicles can endanger passengers in all connected vehicles. Moreover, the attacker can evade privacy by leaking personal information such as vehicle identity, current vehicle position, speed, and acceleration.

> **Open Question 2:** How can the robot immediately detect and respond to a security breach? Can the robot alert the administrator about the intruder?

– *Mutual Authentication:* Authentication has become one of the main concerns in robotics. Mutual authentication is necessary to establish secure communication between robots and humans. Several works have been done to authenticate users, such as face recognition, voice recognition [52], behavior-based recognition [3] etc. However, as we are employing an increasing number of robots in our work, the robots' identities need to be verified as well. Some delivery robots [26,43,48] use OTP (One-Time Password) or mobile applications on users' smartphones to authenticate to the user. But these methods are insufficient because they are susceptible to attacks [36]. Adi et al. proposed an *unclonable identity* for robots based on the work [1]. This identity will be unique to human DNA. However, this process is complex, expensive, and not feasible for mass production. Later, Gavrilova et al. [16] presented an idea to use biometric principles (e.g., physical and behavioral characteristics) to recognize and authenticate *virtual* avatars.

> **Open Question 3:** Is it possible to assign unique biometrics for robot authentication?

– *Autonomous Robot:* The current generation of robots is not fully autonomous; they depend on pre-programmed commands. However, several initiatives are underway to extend the perimeter and allow robots to have autonomy to some extent, e.g., unmanned vehicles, Tesla bot [49].

Military services are also trying to utilize autonomous robots in war, spying, bomb defusal, and other dangerous jobs. However, the use of robots at war is a controversial topic, as it can violate international *Humanitarian law* [47]. The question arises with the *Robot at war*, what happens when an order contradicts the *war robot*'s system. For example, if a robot receives an order to attack a house, the robot detects with sensors that the house is full of children. The order contradicts the robot's system in minimizing civilian casualties. Should the robot be allowed to have an awareness of these types of situations, or should the order override the robot's system [30]?

> **Open Question 4:** What if autonomous robots start to make decisions or refuse orders that might cause harm to humans, like kicking back a human who kicks it?

- **Robot Learning:** Robot Learning [12] is popular for teaching robots without programming every movement explicitly. Robots can learn from demonstrations, teleoperations, or observation [27]. Learning methods can be supervised, unsupervised, transfer learning, and reinforcement learning [41]. The robots adapt their decisions as they perceive the environment or dataset. The attackers can intentionally manipulate the data during the learning process, such as injecting poisonous data into the training set, spoofing sensor data (e.g., camera, audio), or changing learning conditions. Due to these attacks, robots may learn unsolicited behaviors that can exhibit danger to their surroundings. For example, Yang et al. [55] demonstrated an adversarial attack on a reinforcement learning-based robot learning system where the attacker uses a pulse to generate random observations, degrading the learning performance.

> **Open Question 5:** How can anomalies in robot training data be discovered and addressed so that the robot does not learn and perpetuate dangerous behavior?

- **Integration with ChatGPT:** Robots are expected to undergo revolutionary changes using ChatGPT, especially ChatGPT-4. We have seen some proposed frameworks [19,51] in recent times. Vemprala et al. [51] suggested using a ChatGPT prompt to write code automatically for non-technical users to make the robot perform a certain task. In one scenario, the user asks the robot to cook an omelet and serves it to the user's grandfather. Recently, Google DeepMind introduced Robotic Transformer 2 (RT-2), a novel vision-language-action (VLA) model that learns from web-scale datasets [56]. This model is built on the same tech as ChatGPT; It can interpret these data as plain language instruction and execute it [14].

> **Open Question 6:** If ChatGPT can be successfully implemented on robots, what if robots can write code and modify themselves in an unwanted way?

- **Access Control:** Certain robots (e.g., service robots in our homes) continuously surveil us as part of their functions. These robots have access to our personal data; they can take pictures and videos, and monitor our locations. Nonetheless, if the vendor of these robots unethically grants access to the robots' system during manufacturing and takes advantage of our confidential data, it can pose significant privacy and security risks. For example, unauthorized users can collect passwords and credit card information by simply taking photos or videos when the user is entering the data.

> **Open Question 7:** How can we effectively incorporate access control in robots to protect the security and privacy of the end users?

- ***Trolley Problem in Robotics:*** Imagine a scenario where a person is watching a runaway trolley heading towards a track where five people are standing, and if nothing is done, these people will *certainly* die. There is another track where he can divert the trolley, but there is another person standing on it that will be killed. Here arises the ethical dilemma of whether killing one person is okay instead of killing five people. As robots become more involved in society, they will inevitably encounter many ethical dilemmas in decision-making. So, it is essential to solve the trolley problem to mitigate any risks that an action of the robot may pose.

> ***Open Question 8:*** What would be the robot's reaction during a 'Trolley Problem' [24] scenario?

4   Conclusion

The widespread adoption of robots signals the imminent revolution of robotics technology. It may not be very long before we generalize the idea of coexisting with robots. We must be prepared for the privacy and security risks to embrace this transition fully. Robotic systems are made of different subsystems and subcomponents. Securing the subcomponents is necessary but not sufficient for protecting the whole system. This is because components are integrated with one another and therefore, exhibit complex and subtle dependencies and interactions [35]. We need to enforce a robotics framework and a universal policy for developing or changing any robots. Such a comprehensive measure will ensure that robots and their manufacturer follow the standard user safety practice. European Commission has created a *voluntary* code of ethics and standards for manufacturers and users of robotics technology [37]. IEEE undertakes a global initiative- *The IEEE Global Initiative on Ethics of Autonomous and Intelligent Systems*, which aims to ensure that the involved persons prioritize ethical consideration and benefits of humankind [20]. However, as these policies are not enforced as obligatory, the concerns still prevail.

References

1. Adi, W.: Clone-resistant DNA-like secured dynamic identity. In: 2008 Bio-inspired, Learning and Intelligent Systems for Security, pp. 148–153 (2008). https://doi.org/10.1109/BLISS.2008.33
2. Agrawal, S., Williams, M.A.: Robot authority and human obedience: a study of human behaviour using a robot security guard. In: Proceedings of the Companion of the 2017 ACM/IEEE International Conference on Human-Robot Interaction, pp. 57–58 (2017)
3. Almohamade, S.S., Clark, J.A., Law, J.: Behaviour-based biometrics for continuous user authentication to industrial collaborative robots. In: Maimut, D., Oprina, A.-G., Sauveron, D. (eds.) SecITC 2020. LNCS, vol. 12596, pp. 185–197. Springer, Cham (2021). https://doi.org/10.1007/978-3-030-69255-1_12

4. Amoozadeh, M., et al.: Security vulnerabilities of connected vehicle streams and their impact on cooperative driving. IEEE Commun. Mag. **53**(6), 126–132 (2015)
5. Bernadotte, A.: Cyber security for surgical remote intelligent robotic systems. In: 2023 9th International Conference on Automation, Robotics and Applications (ICARA), pp. 65–69 (2023). https://doi.org/10.1109/ICARA56516.2023.10126050
6. Bisconti Lucidi, P., Nardi, D.: Companion robots: the hallucinatory danger of human-robot interactions. In: Proceedings of the 2018 AAAI/ACM Conference on AI, Ethics, and Society, pp. 17–22 (2018)
7. Bonaci, T., Herron, J., Yusuf, T., Yan, J., Kohno, T., Chizeck, H.J.: To make a robot secure: An experimental analysis of cyber security threats against teleoperated surgical robots (2015). arXiv preprint arXiv:1504.04339
8. Booth, S., Tompkin, J., Pfister, H., Waldo, J., Gajos, K., Nagpal, R.: Piggybacking robots: Human-robot overtrust in university dormitory security. In: Proceedings of the 2017 ACM/IEEE International Conference on Human-Robot Interaction, pp. 426–434 (2017)
9. Bristeau, P.J., Dorveaux, E., Vissière, D., Petit, N.: Hardware and software architecture for state estimation on an experimental low-cost small-scaled helicopter. Control. Eng. Pract. **18**(7), 733–746 (2010)
10. Choi, H., et al.: Detecting attacks against robotic vehicles: a control invariant approach. In: Proceedings of the 2018 ACM SIGSAC Conference on Computer and Communications Security, pp. 801–816 (2018)
11. Clark, G.W., Doran, M.V., Andel, T.R.: Cybersecurity issues in robotics. In: 2017 IEEE Conference on Cognitive and Computational Aspects of Situation Management (CogSIMA), pp. 1–5. IEEE (2017)
12. Connell, J.H., Mahadevan, S.: Robot learning. In: Sammut, C., Webb, G.I. (eds.) Encyclopedia of Machine Learning. Springer, Boston, MA (2011). https://doi.org/10.1007/978-0-387-30164-8_732
13. Dash, P., Karimibiuki, M., Pattabiraman, K.: Out of control: stealthy attacks against robotic vehicles protected by control-based techniques. In: Proceedings of the 35th Annual Computer Security Applications Conference, pp. 660–672 (2019)
14. Edwards, B.: Google's RT-2 AI model brings us one step closer to WALL-E (2023). https://arstechnica.com/information-technology/2023/07/googles-rt-2-ai-model-brings-us-one-step-closer-to-wall-e/
15. Eykholt, K., et al.: Robust physical-world attacks on deep learning visual classification. In: Proceedings of the IEEE Conference on Computer Vision and Pattern Recognition, pp. 1625–1634 (2018)
16. Gavrilova, M.L., Yampolskiy, R.V.: Applying biometric principles to avatar recognition. In: 2010 International Conference on Cyberworlds, pp. 179–186 (2010). https://doi.org/10.1109/CW.2010.36
17. van Genderen, R.H.: Privacy and data protection in the age of pervasive technologies in AI and robotics. Eur. Data Prot. Law Rev. **3**, 338–352 (2017). https://doi.org/10.21552/edpl/2017/3/8
18. Gibbs, S.: Hackers can hijack Wi-Fi Hello Barbie to spy on your children) (2015). https://www.theguardian.com/technology/2015/nov/26/hackers-can-hijack-wi-fi-hello-barbie-to-spy-on-your-children
19. He, H.M.: RobotGPT: From chatGPT to robot intelligence (2023). https://openreview.net/forum?id=wWe_OqpCcU8
20. IEEE: The IEEE Global Initiative on Ethics of Autonomous and Intelligent Systems (2017). https://standards.ieee.org/wp-content/uploads/import/documents/other/ec_about_us.pdf

21. Injected, Demos, D.: DoS attacks on a self-balancing robot (accelerometer) (2018). https://youtu.be/yDz8y_ht3Xg
22. Injected, Demos, D.: Side-Swing attacks on a self-balancing robot (2018). https://youtu.be/oy3B1X41u5s
23. International Federation of Robotics (IFR): Service Robots as Defined by ISO 8373. https://ifr.org/service-robots
24. Kamm, F.M.: The Trolley Problem Mysteries. Oxford University Press (2015)
25. Kirschgens, L.A., Ugarte, I.Z., Uriarte, E.G., Rosas, A.M., Vilches, V.M.: Robot hazards: from safety to security (2018). arXiv preprint arXiv:1806.06681
26. Kiwibot: Kiwibot. https://www.kiwibot.com/
27. Kroemer, O., Niekum, S., Konidaris, G.: A review of robot learning for manipulation: challenges, representations, and algorithms. J. Mach. Learn. Res. **22**(1), 1395–1476 (2021)
28. Lacava, G., et al.: Cybser security issues in robotics. J. Wirel. Mob. Netw. Ubiquitous Comput. Dependable Appl. **12**(3), 1–28 (2021)
29. Larson, S.: Ransomware experiment shows the dangers of hacking robots (2018). https://money.cnn.com/2018/03/09/technology/robots-ransomware/index.html
30. Lin, P., Bekey, G.A., Abney, K.: Robots in war: issues of risk and ethics (2009)
31. Liu, Y., Ma, X., Bailey, J., Lu, F.: Reflection backdoor: a natural backdoor attack on deep neural networks. In: Vedaldi, A., Bischof, H., Brox, T., Frahm, J.-M. (eds.) ECCV 2020. LNCS, vol. 12355, pp. 182–199. Springer, Cham (2020). https://doi.org/10.1007/978-3-030-58607-2_11
32. Luo, M., Myers, A.C., Suh, G.E.: Stealthy tracking of autonomous vehicles with cache side channels. In: 29th USENIX Security Symposium (USENIX Security 20), pp. 859–876 (2020)
33. Lutz, C., Schöttler, M., Hoffmann, C.P.: The privacy implications of social robots: scoping review and expert interviews. Mob. Media Commun. **7**(3), 412–434 (2019)
34. Mayoral-Vilches, V., Carbajo, U.A., Gil-Uriarte, E.: Industrial robot ransomware: Akerbeltz. In: 2020 Fourth IEEE International Conference on Robotic Computing (IRC), pp. 432–435 (2020). https://doi.org/10.1109/IRC.2020.00080
35. McDaniel, P., Koushanfar, F.: Secure and trustworthy computing 2.0 vision statement (2023). arXiv preprint arXiv:2308.00623
36. Mulliner, C., Borgaonkar, R., Stewin, P., Seifert, J.-P.: SMS-based one-time passwords: attacks and defense. In: Rieck, K., Stewin, P., Seifert, J.-P. (eds.) DIMVA 2013. LNCS, vol. 7967, pp. 150–159. Springer, Heidelberg (2013). https://doi.org/10.1007/978-3-642-39235-1_9
37. Nevejans, N.: EUROPEAN CIVIL LAW RULES IN ROBOTICS (2016). http://www.europarl.europa.eu/committees/fr/supporting-analyses-search.html
38. Niemeyer, G., Preusche, C., Stramigioli, S., Lee, D.: Telerobotics. In: Siciliano, B., Khatib, O. (eds.) Springer Handbook of Robotics, pp. 1085–1108. Springer, Cham (2016). https://doi.org/10.1007/978-3-319-32552-1_43
39. Othman, F., Bahrin, M., Azli, N., et al.: Industry 4.0: a review on industrial automation and robotic. J. Teknol. **78**(6–13), 137–143 (2016)
40. Petit, J., Shladover, S.E.: Potential cyberattacks on automated vehicles. IEEE Trans. Intell. Transp. Syst. **16**(2), 546–556 (2015). https://doi.org/10.1109/TITS.2014.2342271
41. Ranaweera, M., Mahmoud, Q.H.: Virtual to real-world transfer learning: a systematic review. Electronics **10**(12), 1491 (2021)
42. REUTER, E.: Hospitals sue surgical robot maker, saying it forced them into restrictive contracts. https://medcitynews.com/2021/07/hospitals-sue-surgical-robot-maker-saying-it-forced-them-into-restrictive-contracts/ (2021)

43. Serve: Serve Robotics Becomes First Autonomous Vehicle Company to Commercially Launch Level 4 Self-Driving Robots. https://www.serverobotics.com/level-4-autonomy
44. Shah, R., Ahmed, M., Nagaraja, S.: Fingerprinting robot movements via acoustic side channel (2022). arXiv preprint arXiv:2209.10240
45. Shah, R., Ahmed, M., Nagaraja, S.: Reconstructing robot operations via radio-frequency side-channel (2022). arXiv preprint arXiv:2209.10179
46. Sugawara, T., Cyr, B., Rampazzi, S., Genkin, D., Fu, K.: Light commands: Laser-Based audio injection attacks on Voice-Controllable systems. In: 29th USENIX Security Symposium (USENIX Security 20), pp. 2631–2648. USENIX Association (2020). https://www.usenix.org/conference/usenixsecurity20/presentation/sugawara
47. Szpak, A.: Legality of use and challenges of new technologies in warfare - the use of autonomous weapons in contemporary or future wars. Eur. Rev. **28**(1), 118–131 (2020). https://doi.org/10.1017/S1062798719000310
48. Team, Y.S.D.: The story behind the creation of Yandex's delivery robot (2021). https://medium.com/yandex-self-driving-car/the-story-behind-the-creation-of-yandexs-delivery-robot-e07017940589
49. Tesla: Tesla Bot Update (2023). https://www.youtube.com/watch?v=XiQkeWOFwmk
50. Tu, Y., Lin, Z., Lee, I., Hei, X.: Injected and delivered: fabricating implicit control over actuation systems by spoofing inertial sensors. In: 27th USENIX Security Symposium (USENIX Security 18), pp. 1545–1562 (2018)
51. Vemprala, S., Bonatti, R., Bucker, A., Kapoor, A.: ChatGPT for robotics: design principles and model abilities. Microsoft Auton. Syst. Robot. Res **2**, 20 (2023)
52. Wang, T.M., Tao, Y., Liu, H.: Current researches and future development trend of intelligent robot: a review. Int. J. Autom. Comput. **15**(5), 525–546 (2018)
53. Winfield, A.F.T., Winkle, K., Webb, H., Lyngs, U., Jirotka, M., Macrae, C.: Robot accident investigation: a case study in responsible robotics. In: Software Engineering for Robotics, pp. 165–187. Springer, Cham (2021). https://doi.org/10.1007/978-3-030-66494-7_6
54. Yaacoub, J.P.A., Noura, H.N., Salman, O., Chehab, A.: Robotics cyber security: vulnerabilities, attacks, countermeasures, and recommendations. Int. J. Inf. Secur. 1–44 (2022)
55. Yang, C.H.H., et al.: Enhanced adversarial strategically-timed attacks against deep reinforcement learning. In: ICASSP 2020–2020 IEEE International Conference on Acoustics, Speech and Signal Processing (ICASSP), pp. 3407–3411. IEEE (2020)
56. Yevgen Chebotar, T.Y.: RT-2: New model translates vision and language into action. https://www.deepmind.com/blog/rt-2-new-model-translates-vision-and-language-into-action (2023)
57. Yogeeswaran, K., Złotowski, J., Livingstone, M., Bartneck, C., Sumioka, H., Ishiguro, H.: The interactive effects of robot anthropomorphism and robot ability on perceived threat and support for robotics research. J. Hum.-Robot Interact. **5**(2), 29–47 (2016)
58. Zhu, Q., Rass, S., Dieber, B., Vilches, V.M., et al.: Cybersecurity in robotics: challenges, quantitative modeling, and practice. Found. Trends® Robot. **9**(1), 1–129 (2021)

Waves of Knowledge: A Comparative Study of Electromagnetic and Power Side-Channel Monitoring in Embedded Systems

Michael Amar[1]([✉]), Lojenaa Navanesan[1,2], Asanka P. Sayakkara[2], and Yossi Oren[1]

[1] Ben -Gurion University of the Negev, Beer Shea, Israel
amarmic@post.bgu.ac.il, {lojenaa,yos}@bgu.ac.il
[2] University of Colombo School of Computing, Colombo, Sri Lanka
asa@ucsc.cmb.ac.lk

Abstract. In today's interconnected world, Programmable Logic Controller (PLC) devices play a crucial role in controlling and automating critical processes across various sectors. This increased connectivity, however, also brings about significant security risks, including the threat of the PLC's control flow being subverted through malicious code injected by state-level actors. This paper offers an exploration of the use of side channels for control flow monitoring. By analyzing subtle variations in system behavior, such as power consumption and electromagnetic radiation, these side channels can be effectively leveraged to infer control flow information, and thus identify potential attacks. To accomplish this, we employ the emitted signals to train a machine learning model, and evaluate our detector by simulating two different types of attacks: malicious code injection and sensitive data infiltration. Additionally, we provide a unique comparison between the power consumption and electromagnetic side channels, highlighting the primary benefits each signal type exhibits in terms of detecting and preventing attacks. The results presented in this paper can aid system manufacturers in selecting the most suitable channel for defending their system, based on the specific requirements and context of their PLC application.

Keywords: Physical side-channel analysis · Malware detection · Malware monitoring · PLC environment · Firmware verification

1 Introduction

The rise in use of cyber-physical systems (CPS), which include smart vehicles, industrial systems, medical monitoring, robotics, and more, promises to modernize society and to reduce the burden of human labor [5, 7]. They typically consist

---

Michael Amar, Lojenaa Navanesan, Both authors are considered co-first authors.

© ICST Institute for Computer Sciences, Social Informatics and Telecommunications Engineering 2024
Published by Springer Nature Switzerland AG 2024. All Rights Reserved
Y. Chen et al. (Eds.): SmartSP 2023, LNICST 552, pp. 158–170, 2024.
https://doi.org/10.1007/978-3-031-51630-6_11

of a large-scale, interconnected system of disparate elements that integrate computation with physical processes. CPS may significantly increase the effectiveness of industrial process control systems (ICS) [19]. Programmable logic controllers (PLCs) are a vital component of the CPS. They are used by industrial control systems (ICS) to link to and monitor critical infrastructure, employed in the manufacturing and process industries to reduce costs and enhance quality, and were created in response to the requirement to replace traditional programmed relay panels [8]. PLCs are designed for specific tasks combining multiple functionality in large industries. In recent times, PLCs have largely taken the place of the control components that were formerly used to execute the logic of the system [3]. PLCs gather data and communicate with sensors, motors, valves, and other equipment positioned throughout massive industrial systems to automate and control manufacturing processes. PLCs are *operational devices* – they are directly connected to the physical system in supervisory control and data acquisition (SCADA) in operational technology (OT) and information technology (IT) [12]. The PLCs themselves are typically observed using a remote human-machine interface [6, 16].

The increasing reliance on industrial control systems (ICSs) has made them an attractive target for attackers looking to disrupt operations or gain unauthorized access to sensitive information. PLCs are particularly susceptible to attacks because of their widespread usage and lack of built-in security features. An attacker who gains control of a PLC can directly interfere with the underlying industrial processor and influence its interactions with the physical world. An attacker may also use this control of the PLC to exfiltrate secret data, such as sensor readings, which are exposed to the PLC. One of the most famous attacks on PLCs is the Stuxnet virus, that tampered with the code running on a PLC and disrupted the Iranian nuclear program by changing the rotational speed of the centrifuges [13]. Indeed, preventing attacks on the PLC environment is a very important task. While traditional security measures, such as firewalls, intrusion detection systems, and encryption, can help protect PLC environments they are not foolproof. As a result, many works have suggested PLC-specific attack deterrence and prevention measures [4, 20, 21]. One measure which has been suggested for detecting and preventing malware attacks on PLC environments is the use of non-intrusive passive integrity monitors. Abadi et al. [2] define Control Flow Integrity as "*security policy dictates that software execution must follow a path of a Control-Flow Graph (CFG) determined ahead of time*"; ensuring this policy can protect software from control flow hijacking attacks caused by buffer overflow, code reuse, or similar attacks. One approach for building this monitor is through the use of power or electromagnetic (EM) radiation side channels. An anomalous signal would imply a deviation from the predefined CFG, and can be an indication for an attack. Since the side channel-based monitor is air-gapped from the rest of the PLC by design, it is not susceptible to the same attack vectors as the PLC, thereby preventing any potential attacks on the monitor through similar channels.

Common approaches for passive code monitoring are the power and EM side channels. Both approaches have advantages and disadvantages, and were

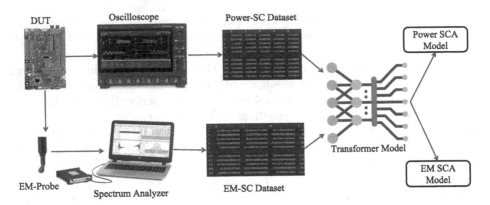

Fig. 1. *The process of acquiring power and electromagnetic side-channel traces from the target device and applying the deep-learning techniques to generate individual models for anomaly detection.*

suggested in several works previously, but they were never compared directly in a PLC setting. Liu et al. [14] were able to recover the program execution flow by observing the power consumption of a microcontroller; they inferred what instruction is most likely executed with an improved hidden Markov model. Han et al. [12] presented a non-intrusive EM based monitor, and trained an LSTM-based detector for signals in the time and frequency domains. We adopt some of their ideas as foundational concepts in our study.

In this paper, we investigate the suitability of physical side-channel analysis to monitoring and preventing attacks in PLC environments. Our approach leverages advanced signal processing and machine learning techniques to identify anomalous behavior in the physical signals produced by PLCs and detect potential attacks. We compare the power and EM side-channel approaches to identify the most effective side-channel medium to prevent and monitor attacks on the PLC environment. We leverage both EM and power consumption side channels to profile the behavior of our Device Under Test (DUT), and train a machine learning model based on the acquired signals. A general overview of our experimental environment can be seen in Fig. 1.

The contributions of this paper are as follows:

- We provide an anomaly detector to identify attacks against embedded controllers in time critical environments. We simulate two types of attacks, and evaluate our detector's effectiveness in detecting them on a popular controller.
- We present a transformer-based model architecture that is agnostic to the type of data used as input. The architecture is suitable for both EM and power consumption signals.
- We perform a comparison between the EM and power side channels, and provide criteria for choosing between them, considering the constraints of the PLC environment.

Our research, comparing the EM and power side channels, yields valuable insights and guidance for decision-makers to strategically select the optimal monitoring approach tailored to their unique environmental conditions. By leveraging this research, organizations can confidently implement robust security measures, fortifying their systems against potential attacks and safeguarding critical infrastructure with heightened resilience and effectiveness.

1.1   Background

The Power and EM Side Channels. A side channel can be defined as a medium through which sensitive information can be inadvertently revealed during the operation of a system. This medium can take various forms, such as power consumption, EM radiation, timing, or sound. Attackers can exploit side channels to extract sensitive information, such as cryptographic keys or passwords, without direct access to the system's memory [10]. From the defender's perspective, side channels can provide insight into the code that is currently being executed, allowing them to ensure the system's reliability.

In modern processors, transistors are continually switching on and off, causing a varying current and resistance in the digital circuit. In addition to this direct effect on power consumption, any metallic substance in proximity to the circuit acts as an antenna and transmits an electromagnetic wave in response. Usually the range of the EM waves is limited, and an amplifier is needed to enhance the strength of the signal. The shape and characteristics of the signals produced by these side channels are influenced by two main factors: the executed instructions and the processed data [15].

This effect makes it possible to gain insights about the instructions and the data by observing the side-channel trace, making this method a natural candidate for anomaly detection based on control flow monitoring. To demonstrate the effectiveness of side-channel measurements in determining which code is currently being executed, we ran four different applications (AES encryption, Matrix multiplication, Random number generator, Idle program) on our DUT, collected the resulting EM signals and plotted the first 3 PCA coefficients of these signals on a grid. As Fig. 2 shows, it is apparent that each application forms its own cluster, emphasizing the distinct waves emitted by the DUT during execution. As this pilot experiment illustrates, side-channel signals can clearly be analyzed to infer information about the code being executed and the data being processed.

Transformer Networks. Transformers are a type of neural network architecture that has gained popularity in recent years by improving the performance of natural language processing (NLP) tasks. They were first introduced in 2017 by Vaswani et al. [18] as an approach to machine translation, and have since then transformed the field of NLP. Previous approaches to NLP tasks involved the use of recurrent neural networks (RNNs) or convolutional neural networks (CNNs) to process sequences of words. However, these models have limitations in capturing long-range dependencies and suffer from the vanishing and exploding

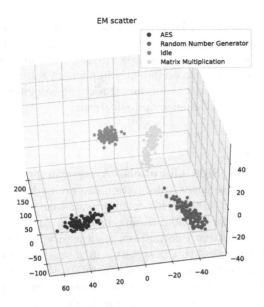

Fig. 2. *Scatter of four applications after performing PCA on their EM signals.*

gradient problems. Transformers are designed to address these issues by utilizing a self-attention mechanism that allows them to capture long-range dependencies without the need for recurrence. The self-attention mechanism in transformers allows the model to attend to different parts of the input sequence while processing each element. This is achieved through the use of attention weights that determine the importance of each input element to the output. One of the key benefits of the transformer architecture is its ability to process input sequences in parallel, making it much faster than traditional RNNs like LSTMs. LSTMs may, however, suffer from vanishing and exploding gradient problem, and their parallelization potential is limited since each time step depends on the previous one. The detection technique in this paper is based on the work of Han et al. [12]. They employed an LSTM-based model, which utilized a hidden state vector to represent the unobserved code, while the observables were the EM signals.

2   Methods

In order to establish a benchmark for comparing EM and power side channels, we developed an agnostic detector that identifies the executed code sections and also detects anomalies during runtime. Subsequently, we evaluated it using both EM and power signals. The results presented in this paper can serve as a guide for selecting the suitable monitoring medium based on the specific environmental conditions of the operator.

   To demonstrate this idea, we used the Traffic Alert and Collision Avoidance System (TCAS) program provided by Han et al. [11]. The program is written in C and meant to prevent midair collisions between airplanes. It receives as input information about the position and status of its own and approaching airplanes. We modified the code by adding function calls which use the general purpose input output (GPIO) pins of the DUT to signal the measurement tools the start and end of each *scan cycle*. Scan cycle is the term used to describe the repetitive manner in which PLCs operate: they execute a single program in an infinite loop which reads sensor inputs (e.g. water level sensor), performs control logic which defines the relations between the input and output values, then updates the actuators (e.g. activate a pump). Since the control logic does not change, and since the characteristics of the side signals are mainly influenced by the instructions and processed data, similar input values should yield similar side-channel emissions. In practice, however, a single program may have multiple control flows. Thus, different input values may cause different instructions to be executed, resulting in different side-channel emissions. To train a robust anomaly detector, there is a need for a diverse dataset which contains traces that represent multiple control flows of the program.

To accomplish that, Han et al. [11] fed the source code into the KLEE symbolic execution engine, a static analysis tool that produced multiple sets of variable assignments, where each set leads to a different execution path. For example, Fig. 3 shows a transition from the source code into a CFG. To follow the path $1 \rightarrow 2 \rightarrow 3 \rightarrow 5 \rightarrow 7 \rightarrow 9$ in that CFG, certain conditions have to be satisfied: $low \leq high$ and $x = v[mid]$. This set of constraints was next fed into a Satisfiability Modulo Theories (SMT) solver (e.g. Z3) which returns the exact values of the variables to follow the desired path. This process can be repeated multiple times to get full path coverage.

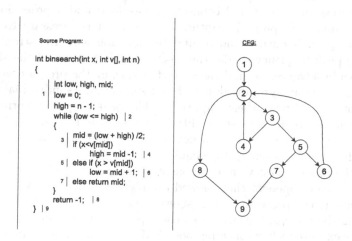

Fig. 3. *Transition from source code to Control flow Graph.*

After generating the test cases, we executed each test case while collecting the signals the controller emitted with dedicated equipment. The power and EM traces were collected simultaneously, to ensure similar operating conditions. We examined a total of 24 test cases representing different logic control flows.

The acquired signals form a dataset, where each sample is a side channel signal, and each label is the execution path that was executed to create it. This dataset was used to train a transformer-based classifier. If an attacker were to modify the control logic or hijack the control flow, the emitted signals would deviate from any known execution. In such case, the confidence of the classifier for all the possible control flows would be low. Thus, if the maximum confidence of the classifier over all possible control flows is lower than a pre-defined threshold, this should trigger an alert. The benefit of this approach is that the model can both observe the currently executed code and detect malicious execution. Additionally, this approach doesn't require any malicious samples for the training process.

We used a Nordic NRF52-DK as our DUT. This system is equipped with an ARM Cortex M4 processor running at 64 MHz with 64 KB RAM, and is designed for the I/O and digital signal control markets [1]. The DUT was connected to a Keysight B2962A low noise power source. For power consumption measurements, we used a Keysight MSOS604A Oscilloscope with 1GSa/s with High Resolution mode. We post-processed the power traces by averaging every 16 samples to reduce noise. For EM measurements, we used a Tektronix RSA306 Spectrum Analyzer together with a LANGER LF-U5 probe and PA 303 amplifier at a sampling rate of 56MS/s.

Measuring power consumption and EM emissions requires a preparation phase. To perform power measurements, the digital circuit needs to be modified by cutting the circuit and connecting a resistor and a probe in series to the power supply. To properly capture EM traces, it is necessary to both find the signal's peak frequency and to find the location with the strongest emissions. The peak frequency carries the most valuable features of the signal, and is unknown in advance. To find the correct frequency the program is executed, and the spectrum analyzer divides the input signal into its individual frequency components and displays their amplitudes. The peak frequency corresponds to the frequency with the highest amplitude. To find the location with the most dominant signals, we carried out a device cartography step, in which we used a Secure-IC XYZ positioning stage equipped with SMC100 single axis steppers to move the probe over several components, including the SOC and the MCU, and choose the component that showed the highest amplitude. We identified a location about the processor which showed the most noticeable signals. After performing these steps, the test cases corresponding to the different execution paths were executed while collecting both EM and power signals simultaneously. Each test case was executed 2000 times. The acquired traces are considered as a behavioral baseline of our DUT.

We trained a transformer-based classifier separately for each signal type after performing z-normalization, ensuring the signals are on the same scale. The architecture of the model is inspired by [17], it consists of 8 transformer encoder blocks with 6 attention heads of size 256. Each block consists of two main components: attention and feed-forward. The attention part consists of a normalization layer followed by attention and dropout layers. The output of the attention component is residually connected to the input sequence and then passed to the feed forward layer. The feed forward component contains a convolutional layer followed by a dropout, convolutional and normalization layers. The output of the feed forward component is again residually connected to the output of the attention component. Finally, the output of the encoder blocks is fed into a global average pooling layer and then to a fully connected layer for classification. The model is trained using the AdaMax optimizer. Each model was trained for 10 epochs. We used 80% of the data for training and 20% for testing. The machine learning tasks were implemented in TensorFlow 2.6 for Python 3.9 and run on a cluster containing 58 NVIDIA GeForce GTX 1080 GPUs managed by the Slurm Workload Manager.

3 Results

As noted above, we were motivated to compare the effectiveness of the power and EM monitoring methods. Our models serve two purposes: monitoring the executed code, and identifying deviations from the known control flows. As for the ability of our models to correctly classify each signal to its relevant execution path, both models showed good accuracy. The classification accuracy for the EM and power consumption-based models was 91% and 78%, respectively, meaning that for a given trace, the model was able to classify it to the correct execution path, and infer the instructions sequence that was executed. The confusion matrices of the two classifiers are displayed in Fig. 4. The EM model shows good classification results along with some confusion between specific classes, while the power based model shows confusion between the same classes, along with additional misclassification between other classes. It is important to note that this confusion primarily arises because these classes possess only a few differing instructions, making it challenging to differentiate between them. Nonetheless, this confusion can be considered insignificant, since operational malware would incorporate its unique logic and deviate significantly from the program's known behavior.

To evaluate the ability of the models to detect attacks, we simulated two types of attacks: a code injection attack, and data exfiltration attack. The code injection attack included injecting 5–10 assembly instructions to the source code, modelling an attacker interested in modifying the behavior of the control logic. In the second scenario, the data exfiltration attack, the UART pins of the device were used to leak one byte of sensitive information outside the system. In the presence of a signal that exhibits substantial deviations from the program's known behavior, the classifier's confidence level in all classes would be relatively low. Whenever the confidence level fell below a predefined threshold, we categorized it as an anomaly and initiated an alert.

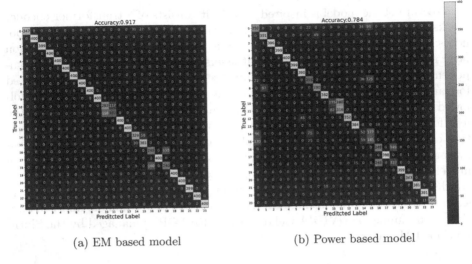

(a) EM based model   (b) Power based model

Fig. 4. Confusion matrices of the models

Both models successfully detected most of the attack scenarios, even the injection of only 5 assembly instructions. AUC reached 97% for the EM model and 91% for the power model. Interestingly, even though the power model was significantly less accurate than the EM model as a classifier, it is only slightly less effective as an anomaly detector. This emphasizes our claim that this approach would prove itself upon a case where the malware chunk has more volume. Garcia et al. [9] assert that the average size of malware, designed with a particular objective for embedded controllers (such as the corruption of the controller's output value), tends to be approximately 2 KB. For that reason, a confusion between the correct control flows which slightly differ from each other, would be negligible as the malware side signals become more dominant.

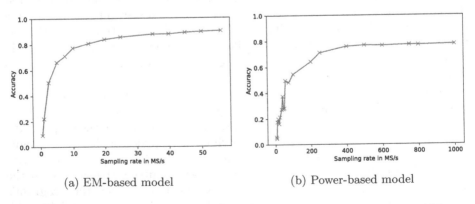

(a) EM-based model   (b) Power-based model

Fig. 5. Effect of the sampling rate on the classification accuracy of the models.

3.1   Performance Analysis

Sampling Rate: Obtaining high-precision equipment like an oscilloscope or a spectrum analyzer is useful to get more detailed analysis with higher sampling rates, but can also incur significant expenses. The sampling rate can impact other aspects, such as the storage and processing requirements for the captured data. Higher sampling rates generate larger amounts of data, which may necessitate more storage capacity and computational resources. We wanted to check the models' performance by exploring the impact of varying sampling rates, and retrained the models accordingly. Figure 5 shows the effect of the sampling rate on the classification accuracy on each signal type. We gradually increased the sampling rate of each signal, as shown on the x-axis. The scale of the x-axis in the graphs differs due to the significantly lower maximum sampling rate of the spectrum analyzer (Fig. 5a) compared to the oscilloscope (Fig. 5b). According to the figure, the EM classifier requires a much lower sampling rate to reach higher accuracy, reaching over 90% accuracy at a sampling rate of 56MS/s, whereas the power model reaches 78% at 1GS/s. Also, while the EM accuracy gradually improves with increasing sampling rate, the power models exhibit a sharp decline around 45MS/s. A possible explanation is that another component of the DUT operates at the same frequency, aliasing the captured signal. This phenomenon is not reflected in the EM classifier, since the measurements are much more localized and are focused only on the component in proximity to the probe.

Noise Resilience: The experiments in this research were conducted under ideal conditions, with minimal electromagnetic interference or environmental noise such as heat or humidity; in practice, however, noise is inevitable. To check the robustness of each model type to noise, we generated white noise with a mean of 0 and varying standard deviation and added it into the z-normalized signals. Figure 6 shows how the noise influences the models' performance. We varied σ, the strength of the noise, from 0 to 9 as shown in the x-axis, the y-axis shows the performance metrics (accuracy and AUC). The dashed orange line shows the performance of the power model while the solid blue line shows the performance of the EM model. When examining the figure, it becomes apparent that the EM model demonstrates superior performance in the presence of minor noise, yet it becomes surpassed as the noise level increases. This makes EM based monitoring the preferred choice for clean environments with minimal interference to the captured signal, where it is possible to place the probe in greater proximity to the measured component, while monitoring through power consumption may be in favor in rougher settings.

4   Discussion

Despite numerous works which have previously suggested side channel monitoring architectures in PLC environments, to the best of our knowledge, there was no comparative study between the EM and power channels for the described topic. This work provides a fair comparison of the power consumption and electromagnetic side channels. Often, due to budgetary and physical constraints, it becomes necessary to determine the appropriate signal type for a given application. The EM signals were much more centralized and precise, as the probe is placed above a component of interest. The model that was trained on the EM signals also showed higher performance metrics (e.g. accuracy, AUC) and required much lower sampling rate to converge, leading to lighter storage and computational requirements. On the other hand, EM capturing requires a preparation phase that includes finding the signal frequency of the examined program and locating the physical position that emits the most distinguishing signals. This process is unique to the DUT used, and must be repeated whenever the hardware configuration of the bill of materials (BOM) changes. The performance of the EM model was also decreased when faced with noisier samples, a realistic risk when taking into account the noisy environments where PLCs are typically found, as well as shielding and extra EM interference that other components of the system may produce. The power model, on the other hand, showed high resilience to noise, which is an important attribute, as often PLCs are exposed extreme environmental conditions. For the downside, measuring power requires a physical modification to the digital circuit which forces it to be shut down, a step which may be impossible in some OT settings. Power traces also include evidence about the consumption of the entire DUT rather than a single component like the processor. Additionally, the power model required much higher sampling rate to converge, resulting in high data volume and heavier processing power.

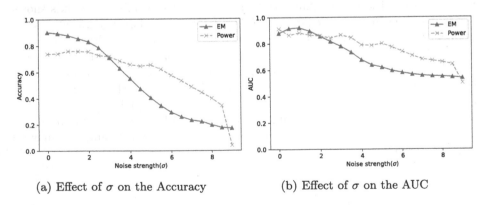

(a) Effect of σ on the Accuracy   (b) Effect of σ on the AUC

Fig. 6. Impact of noise on the models performance.

Detecting anomalies and identifying potential attacks are of paramount importance in ensuring the security of embedded controllers. Leveraging the power of side channels, this paper not only provides effective means for detecting such attacks, but also offers valuable insights to guide the selection of the most appropriate side channel medium, facilitating the design of robust defense mechanisms.

Acknowledgements. This research is supported by the U.S.-Israel Energy Center managed by the Israel-U.S. Binational Industrial Research and Development(BIRD) Foundation.

References

1. Nordic semiconductor. https://www.nordicsemi.com/-/media/Software-and-other-downloads/Product-Briefs/nRF52832-product-brief.pdf
2. Abadi, M., Budiu, M., Erlingsson, Ú., Ligatti, J.: Control-flow integrity principles, implementations, and applications. ACM Trans. Inf. Syst. Secur. **13**(1), 4:1–4:40 (2009). https://doi.org/10.1145/1609956.1609960
3. Alphonsus, E.R., Abdullah, M.O.: A review on the applications of programmable logic controllers (PLCs). Renew. Sustain. Energy Rev. **60**, 1185–1205 (2016). https://doi.org/10.1016/j.rser.2016.01.025, https://www.sciencedirect.com/science/article/pii/S1364032116000551
4. Alves, T., Das, R., Morris, T.: Embedding encryption and machine learning intrusion prevention systems on programmable logic controllers. IEEE Embed. Syst. Lett. **10**(3), 99–102 (2018)
5. Baheti, R., Gill, H.: Cyber-physical systems. Impact Control Technol. **12**(1), 161–166 (2011)
6. Basnight, Z., Butts, J., Lopez, J., Jr., Dube, T.: Analysis of programmable logic controller firmware for threat assessment and forensic investigation. J. Inf. Warfare **12**(2), 1–9 (2013)
7. Bhrugubanda, M.: A review on applications of cyber physical systems. Int. J. Innov. Sci. Eng. Technol. **2**(6), 728–730 (2015)
8. Erickson, K.: Programmable logic controllers. IEEE Potentials **15**(1), 14–17 (1996). https://doi.org/10.1109/45.481370
9. Garcia, L., Brasser, F., Cintuglu, M.H., Sadeghi, A., Mohammed, O.A., Zonouz, S.A.: Hey, my malware knows physics! attacking plcs with physical model aware rootkit. In: 24th Annual Network and Distributed System Security Symposium, NDSS 2017, San Diego, California, USA, February 26 - March 1, 2017. The Internet Society (2017), https://www.ndss-symposium.org/ndss2017/ndss-2017-programme/hey-my-malware-knows-physics-attacking-plcs-physical-model-aware-rootkit/
10. Genkin, D., Shamir, A., Tromer, E.: RSA key extraction via low-bandwidth acoustic cryptanalysis. In: Garay, J.A., Gennaro, R. (eds.) CRYPTO 2014. LNCS, vol. 8616, pp. 444–461. Springer, Heidelberg (2014). https://doi.org/10.1007/978-3-662-44371-2_25
11. Han, Y., Chan, M., Aref, Z., Tippenhauer, N.O., Zonouz, S.: Hiding in plain sight? on the efficacy of power side channel-based control flow monitoring. In: Proceedings of the USENIX Security Symposium (USENIX Security) (2022)
12. Han, Y., Etigowni, S., Liu, H., Zonouz, S.A., Petropulu, A.P.: Watch me, but don't touch me! contactless control flow monitoring via electromagnetic emanations. CoRR **abs/1708.09099** (2017), http://arxiv.org/abs/1708.09099

13. Langner, R.: Stuxnet: Dissecting a cyberwarfare weapon. IEEE Secur. Priv. **9**(3), 49–51 (2011). https://doi.org/10.1109/MSP.2011.67
14. Liu, Y., Wei, L., Zhou, Z., Zhang, K., Xu, W., Xu, Q.: On code execution tracking via power side-channel. In: Proceedings of the 2016 ACM SIGSAC Conference on Computer and Communications Security, pp. 1019–1031 (2016)
15. Mangard, S., Oswald, E., Popp, T.: Power Analysis Attacks. Springer, Boston, MA (2007). https://doi.org/10.1007/978-0-387-38162-6
16. McMinn, L., Butts, J.: A firmware verification tool for programmable logic controllers. In: Butts, J., Shenoi, S. (eds.) ICCIP 2012. IAICT, vol. 390, pp. 59–69. Springer, Heidelberg (2012). https://doi.org/10.1007/978-3-642-35764-0_5
17. Ntakouris, T.: Timeseries classification with a transformer model (2021). https://keras.io/examples/timeseries/timeseries_transformer_classification/
18. Vaswani, A., et al.: Attention is all you need. In: Advances in Neural Information Processing Systems, vol. 30 (2017)
19. Wang, Y., Vuran, M.C., Goddard, S.: Cyber-physical systems in industrial process control. ACM Sigbed Rev. **5**(1), 1–2 (2008)
20. Yılmaz, E.N., Gönen, S.: Attack detection/prevention system against cyber attack in industrial control systems. Comput. Secur. **77**, 94–105 (2018)
21. Ylmaz, E.N., Ciylan, B., Gönen, S., Sindiren, E., Karacayılmaz, G.: Cyber security in industrial control systems: analysis of dos attacks against PLCs and the insider effect. In: 2018 6th International Istanbul Smart Grids and Cities Congress and Fair (ICSG), pp. 81–85. IEEE (2018)

Author Index

Y. Chen et al. (Eds.): SmartSP 2023, LNICST 552, pp. 171–172, 2024.
https://doi.org/10.1007/978-3-031-51630-6

Printed in the United States
by Baker & Taylor Publisher Services